NATO and the North Atlantic

Revitalising Collective Defence

Edited by John Andreas Olsen

www.rusi.org

Royal United Services Institute for Defence and Security Studies

NATO and the North Atlantic: Revitalising Collective Defence
Edited by John Andreas Olsen

First published 2017

Whitehall Papers series

Series Editor: Professor Malcolm Chalmers
Editor: Dr Emma De Angelis

RUSI is a Registered Charity (No. 210639)
ISBN 978-1-138-07961-8

Published on behalf of the Royal United Services Institute for Defence
and Security Studies
by
Routledge Journals, an imprint of Taylor & Francis, 4 Park Square,
Milton Park, Abingdon OX14 4RN

SUBSCRIPTIONS
Please send subscription orders to:

USA/Canada: Taylor & Francis Inc., Journals Department, 530 Walnut Street, Suite 850,
Philadelphia, PA 19106, USA

UK/Rest of World: Routledge Journals, T&F Customer Services, T&F Informa UK Ltd,
Sheepen Place, Colchester, Essex CO3 3LP, UK

Contents

About the Editor

John Andreas Olsen is a colonel in the Royal Norwegian Air Force, currently assigned to London as defence attaché. He is also a visiting professor at the Swedish Defence University and a non-resident senior fellow of the Mitchell Institute. His previous assignments include tours as director of security analyses in the Norwegian MoD, deputy commander at NATO Headquarters, Sarajevo, dean of the Norwegian Defence University College and head of the college's division for strategic studies. Olsen is a graduate of the German Command and Staff College and has served as liaison officer to the German Operational Command and as military assistant to the Norwegian embassy in Berlin. He has a doctorate from De Montfort University and a Master's degree from the University of Warwick. Olsen has published a series of books, including *Strategic Air Power in Desert Storm* (2003), *John Warden and the Renaissance of American Air Power* (2007), *A History of Air Warfare* (2010) and *Airpower Reborn* (2015).

About the Authors

Malcolm Chalmers is deputy director-general of the Royal United Services Institute (RUSI). He was a member of the consultative panel for both the 2010 and 2015 Strategic Defence and Security Reviews and has been a special adviser to Parliament's Joint Committee on the National Security Strategy since 2011. He is honorary visiting professor at the University of Exeter and has been a visiting professor at King's College, London. Before joining RUSI in 2011, Chalmers was a senior special adviser to Foreign Secretaries Jack Straw MP and Margaret Beckett MP. His research focuses on UK defence, foreign and security policy. He is the author of many books and articles, including *Wars in Peace: British Military Operations since 1991* (2014), *Less is Better: Nuclear Restraint at Low Numbers* (2012) and *Uncharted Waters: the UK, Nuclear Weapons and the Scottish Question* (2001).

Heather A Conley is senior vice president for Europe, Eurasia and the Arctic, and director of the Europe Program at the Center for Strategic and International Studies (CSIS). Prior to joining CSIS in 2009, she served as executive director of the Office of the Chairman of the Board at the American Red Cross. From 2001 to 2005, she served as deputy assistant secretary of state in the Bureau for European and Eurasian Affairs with responsibilities for US bilateral relations with the countries of northern and central Europe. From 1994 to 2001, she was a senior associate with an international consulting firm led by former US deputy secretary of state Richard L Armitage. She is a member of the World Economic Forum's Global Agenda Council on the Arctic. She received her BA in international studies from West Virginia Wesleyan College and her MA in international relations from the Johns Hopkins University School of Advanced International Studies (SAIS).

Svein Efjestad became Policy Director at the Norwegian Ministry of Defence (MoD) in 2013. He joined the MoD in 1981 after having completed his master degree in political science from the University of Oslo and a brief period at the Norwegian Institute of International Relations (NUPI). Efjestad has held different positions in the MoD and served at the Norwegian delegation to NATO from 1986 to 1990. From

1995 to 2013 he served as director general for security policy at the MoD and in that capacity he had broad responsibilities in policies regarding national readiness, intelligence, defence planning and NATO affairs. Efjestad has represented the MoD in a number of national and international committees dealing with different security policy issues. In his current position, Efjestad is primarily engaged in policy planning, support to security policy research, the Nordic Defence Cooperation and bilateral defence and security issues. He is also chairman of the Norwegian Coast Guard Council.

John J Hamre was elected president and CEO of CSIS in January 2000. Before joining CSIS, he served as the 26th US deputy secretary of defense. Prior to holding that post, he was the under-secretary of defense (comptroller) from 1993 to 1997 – the principal assistant to the secretary of defense for the preparation, presentation and execution of the defence budget and management improvement programmes. In 2007, Secretary of Defense Robert Gates appointed Hamre as chairman of the Defense Policy Board. Before serving in the Department of Defense, Hamre worked for ten years as a professional staff member of the Senate Armed Services Committee. During that time, he was primarily responsible for the oversight and evaluation of procurement, research, and development programs, defence budget issues and relations with the Senate Appropriations Committee. From 1978 to 1984, John Hamre served in the Congressional Budget Office. He received his PhD from Johns Hopkins SAIS.

Peter Hudson joined the Royal Navy in 1980, specialised as a warfare officer and retired as a Vice Admiral in 2016. He held numerous commands including the T23 ASW Frigate HMS *Norfolk* and was the first Captain of the assault ship HMS *Albion*. Senior seagoing appointments included leading the UK's Amphibious Task Group, serving as a coalition task force commander in the Gulf and, as a Rear Admiral, commanding UK Maritime Forces; an appointment which involved time as the operational commander of EU Naval Forces for Counter Piracy off Somalia and service as NATO's on-call High Readiness Force (Maritime) commander. Ashore, he was Assistant Chief of the Naval Staff (Capability) shaping future capabilities and force development and, on promotion to Vice Admiral in 2013, commanded NATO's Maritime Command at Northwood and was the senior operational maritime advisor to NATO. He is now the Director for International Maritime Programmes with L3 Technologies.

Peter Roberts is Director of the Military Sciences team and a senior research fellow at RUSI. He retired from the Royal Navy in January 2014 after a career as a warfare officer, having commanded a warship and served as a national military representative. His operational experience at sea around the world was broadened by experiences in Bosnia, Iraq and Pakistan in a variety of roles with all three branches of the British armed forces, the US Coast Guard, US Navy, US Marine Corps and intelligence services from several nations. His most recent military experience gave him responsibility for military cyber warfare, information operations, human and signals intelligence, and maritime intelligence, surveillance, target acquisition and reconnaissance collection. Roberts has a Master's degree from King's College, London and a doctorate in History and Politics. He is a visiting lecturer at the University of Portsmouth and a fellow of the Chartered Management Institute.

James Stavridis is a 1976 distinguished graduate of the US Naval Academy who spent more than 35 years on active service in the US Navy. He served for seven years as a four-star admiral, including nearly four years as the first Navy officer chosen to be Supreme Allied Commander of NATO. After retiring from the Navy, he was named dean of The Fletcher School of Law and Diplomacy at Tufts University in 2013. In addition, he currently serves as the US Naval Institute's Chair of the Board of Directors. He has written articles on global security issues for the *New York Times*, the *Washington Post*, *Atlantic Magazine*, *Naval War College Review* and *Proceedings* and he is the author of several books, including *The Accidental Admiral: A Sailor Takes Command at NATO* (2014), *Partnership for the Americas: Western Hemisphere Strategy and U.S. Southern Command* (2010) and *Destroyer Captain: Lessons of a First Command* (2008).

Rolf Tamnes is professor at the Norwegian Institute for Defence Studies (IFS). He served as its director for sixteen years (1996 to 2012), as head of the international research programme 'Geopolitics in the High North' (2008–2012) and adjunct professor at the University of Oslo (1995–2009). Tamnes has been a visiting fellow at CSIS and at St Antony's College, Oxford (2014). He has also been public policy scholar at the Woodrow Wilson International Center for Scholars, Washington, DC and visiting fellow at CSIS (2005–06). Most recently he chaired the Expert Commission on Norwegian Security and Defence Policy, appointed by the Norwegian MoD, and served as a core member in the Afghanistan Inquiry Committee, appointed by the Norwegian government. He has published many books, including co-authored works such as *Common or Divided Security? German and Norwegian Perspectives on Euro-Atlantic Security* (2014), *Geopolitics and Security in the Arctic: Regional Dynamics in a Global World* (2014) and *High North, High Stakes: Security, Energy, Transport, Environment* (2008).

Map 1: Map of the North Atlantic.

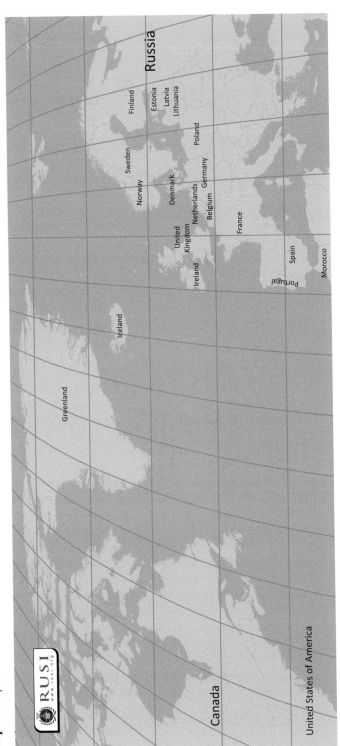

Map 2: Russia's Bastion.

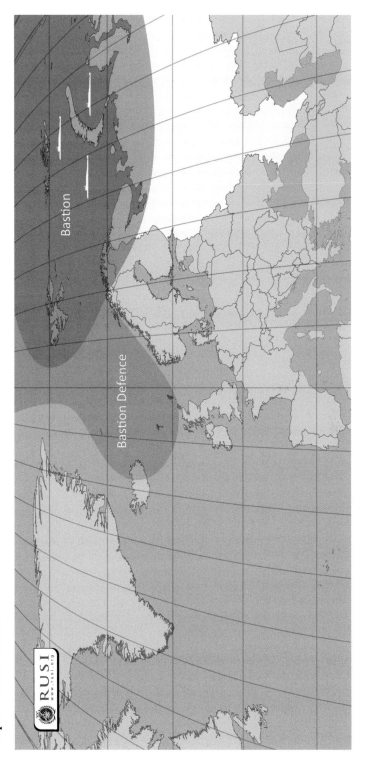

Preface

The North Atlantic Treaty is one of the most important accords of the twentieth century. Signed in Washington, DC, on 4 April 1949 by twelve nations, it formed the basis of the North Atlantic Treaty Organization (NATO), the most successful alliance in recent history. The treaty derives its authority from Article 51 of the United Nations Charter, which articulates the inherent right of independent states to individual or collective defence. The latter concept lies at the heart of the North Atlantic Treaty and is enshrined in Article V, which commits members to protect each other and sets a spirit of solidarity based on common values within the Alliance: an armed attack against one or more shall be considered an attack against them all. While the international security environment has changed and NATO has had to adapt its approach and outlook, its core mission of collective defence remains the same. Despite the accession of new members in the intervening decades NATO has never needed to modify the original fourteen-article treaty. The treaty does not mention any specific threat or adversary; now as in 1949, the members are 'determined to safeguard the freedom, common heritage and civilisation of their peoples, founded on the principles of democracy, individual liberty and the rule of law'.[1]

The North Atlantic – the ocean that connects North America and Europe – is a central part of NATO's area of responsibility. It follows that NATO must retain the capability to secure freedom of manoeuvre across the sea and to keep the waterways between the continents open for reinforcement and resupply of materiel and personnel in times of peace, crisis and war. Thus the Alliance, as the guarantor of national sovereignty and territorial integrity, must be prepared to counter any potential threat to the North Atlantic Ocean. Although this has been the case since the Alliance was formed, it was explicitly acknowledged and reaffirmed at the NATO Summit in Warsaw in July 2016:

> In the North Atlantic, as elsewhere, the Alliance will be ready to deter and defend against any potential threats, including against sea lines of communication and maritime approaches of NATO territory. In this

---

[1] For the full official text, see *The North Atlantic Treaty*.

context, we will further strengthen our maritime posture and comprehensive situational awareness ... The Alliance maritime posture supports the four roles consisting of collective defence and deterrence, crisis management, cooperative security, and maritime security, and thus also contributes to projecting stability. The Standing Naval Forces are a core maritime capability of the Alliance and are the centrepiece of NATO's maritime posture. They are being enhanced and will be aligned with NATO's enhanced NATO Response Force to provide NATO's highest readiness maritime forces. We will continue to reinforce our maritime posture by exploiting the full potential of the Alliance's overall maritime power. Work is under way on the operationalization of the Alliance Maritime Strategy [from 2011], as well as on the future of NATO's maritime operations, which are key to NATO's maritime posture.[2]

NATO's renewed interest in, and commitment to, its transatlantic maritime link is in no small part due to Russia's increasingly provocative rhetoric and behaviour over the past decade. Russia is introducing new classes of conventional and nuclear attack submarines and is modernising its Northern Fleet through the addition of long-range, high-precision missiles. The totality of its modernisation programme adds up to a step-change strengthening of Russian maritime capability in support of an anti-access strategy that could challenge NATO's command of the high seas and, potentially, hold both Europe and North America at existential risk. This represents the re-emergence of a strategy that many in the West assumed had dissipated in the aftermath of the Cold War.

This Whitehall Paper sets out to provide both new and updated perspectives on defence and security related to the North Atlantic by examining the past, present and geostrategic context of NATO's northern region. In doing so, it explores the vital importance of maintaining and sustaining the transatlantic link in today's increasingly complex and multifaceted security environment.

I am grateful to the authors for their contributions. Not only did they meet strict deadlines, but they also participated wholeheartedly in a team effort. I am also grateful to the Royal United Services Institute for publishing this book and to Dr Emma De Angelis and her team of internal and external reviewers. I owe special thanks to Drs Alarik Fritz and H P Willmott and to Captains (US Navy) Sean Liedman and Mark Rudesill for

---

[2] NATO, 'Warsaw Summit Communiqué', press release, 9 July 2016, paras 23, 48, <http://www.nato.int/cps/en/natohq/official_texts_133169.htm>, accessed 30 January 2017.

valuable consultation and discussions. Once again I benefited from Margaret S MacDonald's editorial advice and support. Finally, it must be noted that the opinions and conclusions expressed in this study are those of the editor and authors. They do not represent the official position of any government or institution.

John Andreas Olsen
March 2017

FOREWORD

PHILIP M BREEDLOVE

Europe faces a new and daunting security environment, with significant implications for US national security. Russia is blatantly attempting to change the rules and principles that serve as the foundation of European law and order. The challenge posed by a resurgent Russia is enduring, not temporary, and has consequences far beyond its regional reach. We cannot be certain of what Russia will do next and we cannot fully grasp President Vladimir Putin's intent. What we can do is learn from his actions – and what we see suggests growing Russian capabilities, significant military modernisation and ambitious strategic intent.

The defence of Europe depends on two basic lines of operations. The first is the strength of military power in Europe, including US forces forward stationed, exercised and ready to respond. There is simply no substitute for US forward force presence in Europe. It forms the bedrock of our ability to assure the security of our allies, to deter current and potential adversaries and to respond in a timely manner should deterrence fail. Rotational presence is not an alternative to permanent forward presence; nevertheless, a properly resourced rotational presence can play an important role in meeting the requirements in the European theatre, if it is tightly linked to the permanent forces.

The other line of operation relates to the North Atlantic Ocean and the ability to ensure that US reinforcements can reach Europe in the event of a major confrontation. The North Atlantic is NATO's lifeblood; it is the transatlantic link. After two decades of relative calm, we now see a growing Russian threat to this region. The North Atlantic has become the testing area for Russia's increasingly sophisticated submarines and aircraft. Russia has developed offensive long-range, high-precision capabilities and is building high-end maritime capabilities that could deny NATO members freedom of manoeuvre at sea.

During my period as Supreme Allied Commander Europe from May 2013 to July 2016, my team and I worked hard to strengthen NATO's political cohesion and military posture. Unanimously, we decided to revitalise deterrence and collective defence. NATO demonstrated tremendous unity at the summits in Cardiff in 2014 and Warsaw in 2016. Collectively, NATO members condemned Russia's aggression in Ukraine and its attempt to restore elements of its past empire. We focused our activities primarily on air and land deployment, training, and exercises, and in particular the deployment of forces to the Baltic States and Poland. We strengthened the NATO Reaction Force, established the Very High Readiness Joint Task Force, and witnessed other great initiatives such as the UK-led Joint Expeditionary Force. The Enhanced Forward Presence initiative – boosting NATO's presence in the east and southeast, including the deployment of a US brigade – is more than a signal of commitment. Meanwhile, most NATO members have started to increase their defence budgets.

As we look to the future we need to think more broadly and to re-emphasise the maritime domain. NATO must put the North Atlantic Ocean back on its agenda. We must have command of the sea. The way forward is to strengthen both capability and sustainability, upgrade contingency plans, and reassess the current command and control structure to meet the challenges of tomorrow. NATO needs first-class intelligence and top-notch weapon systems for all domains: air; land; sea; submarine; cyber; and space. NATO's European members also need to improve their readiness and responsiveness so that they can move in a crisis, preferably *before* the shooting starts – at best to deter a war and at worst to get in position to fight before an opponent can shut down the flow of reinforcements.

I am delighted with this study on the North Atlantic and honoured to have been given the opportunity to pen a short foreword. This important and timely book by seasoned experts sounds a call for Allied maritime power and presence in the North Atlantic. There is no single correct answer to what the future holds, but we must prepare as best we can. This excellent book represents an important step in the right direction.

US General (Rtd) Philip M Breedlove
Supreme Allied Commander Europe, NATO, 2013–16

INTRODUCTION: THE QUEST FOR MARITIME SUPREMACY

JOHN ANDREAS OLSEN

Only in the late fifteenth century did Europe come to view the North Atlantic as the route to the rest of the world rather than as the western edge of civilisation. Over the following century, Spain, Portugal, France, England and the Dutch provinces all established themselves as naval powers with overseas empires, with Britain emerging as the pre-eminent global naval power by the middle of the eighteenth century. British domination of the North Atlantic was overwhelming, and by the nineteenth century the trade and economic interests of all Western European powers and of the United States depended on British largesse. This British primacy lasted until the two world wars of the twentieth century and the emergence of the United States as the leading world power. After 1945, the North Atlantic retained its importance as the ocean that connected North America and Europe – a reality recognised in the name of the alliance that links them: the North Atlantic Treaty Organization.

During the Cold War, the North Sea, Norwegian Sea, and the ocean gap between Greenland, Iceland and the United Kingdom (known as the 'GIUK gap') remained essential to the security and prosperity of Europe. It was vital for the Alliance to keep the sea lines of communication between North America and Europe open for transport, supply and reinforcement, and therefore the maritime domain unquestionably fell within NATO's area of responsibility. NATO's maritime capabilities, command structure, large-scale exercises and contingency plans reflected the importance of maintaining maritime superiority in the North Atlantic in general and in the North and Norwegian Seas in particular. The United States took the leadership in this regard, strongly supported by the United Kingdom and in close cooperation with Norway.

The Soviet Union accorded equal priority to the North Atlantic: from the 1960s, the 'bastion' concept, which centred on defending and securing the Soviet sea-based nuclear forces located in the vicinity of the Kola Peninsula,[1] was the Northern Fleet's reason for existence. The Soviet Navy sought control of the Norwegian Sea – covering the vast area between northern Norway and the eastern coast of Greenland, including the Norwegian island of Spitsbergen – and sea-denial down to the GIUK choke points. The Soviet naval expansion conferred greater geostrategic importance upon the north; a war might be won in the Fulda Gap but it might be lost at the GIUK gap.

The North Atlantic represented contested waters until the collapse of the Soviet Union. The new Russia joined the North Atlantic Cooperation Council in 1991 and NATO's Partnership for Peace programme in 1994, and cooperation was further institutionalised through the 1997 NATO–Russia Founding Act. The relationship deepened again in 2002 with the establishment of the NATO–Russia Council as a forum for consultation on security issues and practical cooperation in a wide range of areas. Russia's patrols and other activities at sea and under the oceans all but stopped. Consequently, in 2003 NATO closed its operational headquarters with geographical responsibility for the North Atlantic, and replaced it with a command structure better suited to directing a series of out-of-area operations in the wake of the terrorist attacks of 9/11. NATO's 2010 Strategic Concept confirmed the renewed significance of collective defence and Article V, but explicitly emphasised the significance of two other tasks: crisis management and cooperative security.[2]

Russian Maritime Developments since 2008

As Russia started to regain political, economic and military strength, its leadership took a more confrontational stance not aligned with Western values and the rule of law. Russia's military action in Georgia in August 2008 was the first major indication of events to come. In February and March 2014 Russia used unconventional means to destabilise, illegally occupy and annex Crimea, before sponsoring and encouraging separatist activity in eastern Ukraine. Russia's military actions reshaped the boundaries of a major European state, in violation of international law, and constituted part of a broader strategy of coercion aimed at NATO members and partners. This grand strategic move underpinned President Vladimir Putin's most important foreign policy goals: to restore

---

[1] The Kola Peninsula is located in the far northwest of Russia. It lies almost entirely inside the Arctic Circle and is bordered by the Barents Sea in the north and the White Sea in the east and southeast, offering ice-free waters all year round.

[2] See NATO, 'Strategic Concept: Active Engagement, Modern Defence', November 2010, <http://www.nato.int/cps/en/natohq/topics_82705.htm>, accessed 30 January 2017.

Russia as a respected great power and to establish security buffer zones along its borders.

In parallel, Russia has pursued military reform since 2008 that has seen the structure of its armed forces change. Its officer corps is being professionalised and its capabilities significantly modernised. All three arms of the country's nuclear triad have benefited from investment and recapitalisation programmes. Conventional forces are also being upgraded across the board with an emphasis on two key metrics: firepower and range. Russia's high-end military capability includes new and modernised submarines, better fighter-bomber aircraft, and advanced long-range, high-precision missiles. The new strategic submarines are far quieter than their Cold War predecessors and, like them, are armed with missiles that can strike any location in Europe – capabilities with the potential to cause far-reaching and long-term consequences. Russia's revised Maritime Doctrine of July 2015 places specific emphasis on the High North and on assuring access for its forces into the wider Atlantic.[3]

The former commander of the US Navy's Sixth Fleet, Vice Admiral James G Foggo III, has warned that 'Russian submarines are prowling the Atlantic, testing our defences, confronting our command of the seas, and preparing the complex underwater battlespace to give them an edge in any future conflict'.[4] Vice Admiral Clive Johnstone, Commander of NATO Maritime Command, reports that the Russian increase in submarine activities has once again made the North Atlantic a primary area 'of concern' to the Alliance.[5] Overall, the Russian Federation Navy now operates in areas and at a tempo not seen for almost two decades.

Moscow's increased military activity over the past few years gives reason for concern. The priority Russia has placed on military reform and power projection in a time of economic austerity; the scale and scope of Russia's programmed and 'snap' military exercises in the Arctic Ocean, Barents Sea and Baltic Sea; and the revamping of the Northern Fleet, including the establishment of a new Arctic Command, add urgency to those concerns. Beyond this, Putin seeks to secure Russian access to warm-water ports in the Black Sea and the Mediterranean that could provide logistic support for naval forces deployed to the Atlantic. Stronger and more capable armed forces are preconditions for Russia to re-emerge

---

[3] Steven Horrell, et al., 'Updating NATO's Maritime Strategy', Issue Brief, Atlantic Council, July 2016.

[4] James G Foggo III and Alarik Fritz, 'The Fourth Battle of the Atlantic', *Proceedings Magazine* (Vol. 142/6/1,360, June 2016).

[5] See Jeremy Bender, 'NATO Admiral: We're Seeing More Russian Submarine Activity in the Atlantic than "Since the Days of the Cold War"', *Business Insider*, 3 February 2016.

as a regional great power with global reach, and Russia is steadily shaping its navy to support these aims.[6] Russia has not only engaged in a major upgrade of its sea-based deterrent forces and is strengthening its sea-based anti-ballistic missile systems, but is also investing in strategic-level anti-surface warfare capability.[7] Although the Russian Federation Navy is not on a par with the former Soviet Navy, it is developing high-end strategic capabilities that could potentially disrupt sea operations and project force into the Atlantic Ocean, as well as deny Allied maritime operations in the strategic waters between Greenland, Iceland, the UK and Norway.

Russia has made it a strategic priority to re-establish an offensively oriented navy for operations in the North Atlantic.[8] Putin's offensive stance is very popular with the Russian population; to many he represents the counterweight to the United States and its allies in Europe. Russia is committed to revitalising and updating the bastion concept and this will remain the defining factor for NATO defence planning in the northern region in the foreseeable future. The Kremlin's long-range power-projection strategy presents a major challenge to *all* NATO members and partners. This 'second coming' in the North Atlantic is the new strategic reality for European security: in broader terms, the 'new normal'.

The Purpose of this Whitehall Paper

This Whitehall Paper explores the renewed importance of the North Atlantic Ocean to NATO's security through the lenses of the United States, United Kingdom and Norway in particular. These three NATO members form the territorial rim around the North Atlantic and its peripheral seas. All are maritime nations that have historically taken prime responsibility for security in the region and together with Iceland they form the front line facing a resurgent Russian maritime capability. The US is the most powerful country in the world and the UK the strongest military power in Europe. Norway's long coastline creates an enormous expanse of territorial waters and economic zones, and more than 80 per cent of its ocean areas are located north of the Arctic Circle. Norwegian territorial rights cover an area seven times larger than its mainland territory. Consequently, Norway is a key actor in maintaining peace, stability and security in the north and the

---

[6] Geoffrey Till, 'Future Conditional: Naval Power Sits at the Centre of Russian Strategy', *IHS Jane's Navy International* (Vol. 121, No. 7, September 2016), pp. 10–11.

[7] Geoffrey Till, 'Best and Worst: Preparing Navies for All Possible Worlds', *IHS Jane's Navy International* (Vol. 122, No. 1, January 2017), p. 18.

[8] For a recent assessment, see Kathleen H Hicks et al., *Undersea Warfare in Northern Europe* (Boulder, CO: Center for International and Strategic Studies/ Rowman & Littlefield, 2016). Igor Sutyagin, 'Russia Confronts NATO: Confidence-Destruction Measures', RUSI Briefing Paper, 2016, p. 1.

maritime domain has always played a central role in Norwegian security politics. These three countries, with support from the rest of the northern region, must take the lead to ensure that NATO and its partners devote sufficient resources to this aspect of NATO's area of responsibility.

The subject experts who have contributed to this study address three research questions: *why* is the North Atlantic once again of immediate geostrategic importance; *what* are the enduring and new challenges in this maritime domain; and *how* can NATO members ensure sea-control in the event of a crisis or armed conflict? They seek to decode the 'new normal' by focusing on the strategic developments in the North Atlantic from the early period of the Cold War until the present day. To revitalise NATO's ability to execute collective defence and deterrence against the backdrop of a resurgent Russian navy, they examine NATO's policy, strategy, operations, contingency plans, standing forces, command and control mechanisms, responsiveness, bi- and multilateral cooperation, and exercise and training programmes. They also re-examine pre-positioning arrangements, current initiatives, and future investments in weapon platforms and capabilities.

This Whitehall Paper makes clear that NATO has taken the first step towards re-establishing defence and deterrence through forward bases on its eastern flank and that it is now time for the Alliance to adopt a more comprehensive approach by addressing the maritime domain as well, with an emphasis on the strategically important North Atlantic Ocean. The study extends the discussion, connecting NATO's Eurocentric focus on air and land forces in the Baltic States and Poland with the transatlantic maritime domain, including undersea cables, by exploring political, strategic and operational aspects of defence and security. It offers a holistic approach that underscores the link between Europe and the United States – a connection based on shared values and mutual interests. In short, this paper constitutes a discourse on how to maintain maritime supremacy in an increasingly complex, contested and challenging environment, presenting multiple views of how NATO members can respond both politically and militarily. While many elements of the maritime contest in the North Atlantic bring back memories of the Cold War, the authors make clear that the currrent situation differs greatly from that of the past in both quality and quantity of the forces and armaments involved.

Comparing and contrasting the recent past with the contemporary situation, the final chapter explores the way forward, offering various recommendations. Although the authors have different views on the specifics, the recommendations listed in this chapter represent the common ground. They provide a set menu for responding to a growing threat to the ocean that physically and metaphorically binds North America and Europe together. It is the strength of this bond that underpins the transatlantic commitment to collective defence and deterrence, now as over the past 68 years.

I. THE SIGNIFICANCE OF THE NORTH ATLANTIC AND THE NORWEGIAN CONTRIBUTION

ROLF TAMNES

More than 90 per cent of the world's raw materials, manufactured goods and energy supplies travel by sea. Securing the global commons and sea lines of communication (SLOC) is therefore vital to stability, economic growth and development.[1] In peacetime, securing the sea depends heavily on international law and multinational institutions and regimes, with the 1982 UN Convention on the Law of the Sea as a central point of reference. To achieve the same goal in wartime, the US and NATO would have to rely on sea-control and on their own ability to project power.

The two world wars demonstrated the vital role of the North Atlantic as a strategic connection between North America and Europe and the threat posed by submarines to shipping.[2] With the Soviet naval build-up during the Cold War, especially from the 1960s onwards, securing access for US reinforcements to Europe proved a recurrent concern for NATO.

For a long period after the end of the Cold War, no major threats to the freedom of the seas existed. The West could enjoy what Barry R Posen calls 'command of the commons': worldwide freedom of movement on and under the seas and in the air, with the ability to deny this same freedom

---

[1] Jo Inge Bekkevold and Geoffrey Till (eds), *International Order at Sea: How it is Challenged. How it is Maintained* (London: Palgrave Macmillan, 2016), pp. 3, 342.

[2] Owen R Cote, Jr, *The Third Battle: Innovation in the U.S. Navy's Silent Cold War Struggle with Soviet Submarines,* Newport Papers No. 16 (Newport, RI: Naval War College, 2003).

to enemies.[3] Today, the US and the European members of NATO face challenges to the freedom of the seas. The biggest is the prospect that assertive great powers might use long-range, precision-guided missiles to deny access to littorals. This challenge extends beyond the traditional idea of sea-denial: it represents a broader area denial or anti-access threat. In Europe, Russia under President Vladimir Putin seeks to re-impose a sphere of influence, as well as build capabilities to deny the US the ability to project power. By doing so, he would undermine a key pillar of transatlantic defence cooperation.

The roots of Russia's current policy and strategy go back to the Cold War, when the Soviet Union saw NATO under US leadership as its main adversary. The 'bastion' concept – strategic submarines equipped with ballistic nuclear missiles stationed in northern waters, protected by a defensive perimeter stretching to the Greenland, Iceland and UK (GIUK) gap – is once again vital to Russia's strategy. The options for NATO's response during the Cold War are therefore relevant today.

The first phase of NATO's response in the 1970s was air- and ground-centric; it focused on strengthening land and air capabilities in Europe. Today, NATO's main objective has once again been to build forward presence based mostly on land and air forces. But an alternative approach emerged in the early 1980s with the forward maritime strategy, which prioritised sea-control based power projection. Today's challenges demand a similar, maritime-oriented response.

This chapter outlines the Cold War roots and present features of Russian policy and strategy. It also examines the positions and responses of NATO and Norway both today and during the Cold War, focusing on their efforts to secure the maritime domain. The central thesis is that NATO needs to address the revitalised Russian bastion defence concept and counter the emerging anti-access strategy in the North Atlantic. The chapter offers six considerations that merit special attention for the way forward.

The Soviet Bastion Defence Concept

During the Cold War, NATO often regarded the northern flank as secondary to the central front in Europe. This concealed the fact that the two superpowers increasingly focused on the far north and northern waters. With the advent of intercontinental bombers and missiles in the 1960s, the polar region gained significance as it represented the direct route between the superpowers. The growing importance of the Russian Northern Fleet

---

[3] Barry R Posen, 'Commands of the Commons: The Military Foundation of U.S. Hegemony', *International Security* (Vol. 28, No. 1, Summer 2003), pp. 5–46.

and the bastion defence also meant that the northern flank, the maritime domain and the sea bridge between North America and Europe became a major factor in the Cold War conflict.

The traditional role of the Russian navy had been to support the ground forces as a brown-water service. This began to change in the 1950s. The Northern Fleet took on an increasingly prominent position among the Soviet maritime forces, reflecting the fact that the Kola Peninsula offered ice-free access to the North Atlantic, a central Cold War theatre of operation. The first important change occurred with the Soviet build-up of a large attack submarine force. Initially, in the 1950s, submarines operated off the coast of North America and could put pressure on the Atlantic SLOC. By the late 1950s, the Northern Fleet submarine force had become the largest of the Soviet fleets. In 1980, that fleet alone had 143 attack submarines – more than the entire US Navy. Some 50 per cent of the entire Soviet attack submarine fleet and the most modern major surface vessels now belonged to the Northern Fleet.

In the 1960s, the Soviet Union made significant progress towards becoming a fully fledged maritime power. The navy invested heavily in building a general-purpose blue-water fleet that could project power and support diplomacy around the world – and be better able to deny US access to Europe.

The Soviet Union also launched an ambitious strategic nuclear submarine building programme. By 1980, the Northern Fleet had 47 strategic submarines – about 60 per cent of the total – and also received the most modern and capable submarines. From an operational viewpoint, two developments are of particular interest. The first was the steadily growing importance of strategic submarines in the Soviet nuclear triad, although land-based nuclear systems always retained the leading position. The second important change was the evolution of the bastion defence concept.

Given the limited range of missiles on board early generations of strategic submarines, these boats operated close to the North American coast. This made them very vulnerable during long transits, in part because they had to pass Western sonar barriers. Beginning in the 1970s, the Soviets turned their unfavourable geography to their advantage: the *Delta* class from 1972 and the *Typhoon* class from 1981 carried intercontinental missiles. Based in sanctuaries in the European part of the Arctic Ocean and the Sea of Okhotsk in the Pacific – where they could operate near or under the icecap and be protected by nearby naval and air forces – they could target nearly all the US.[4] This approach differed

---

[4] Jacob Børresen et al., *Norsk forsvarshistorie, bd. 5, Allianseforsvar i endring, 1970–2000* [*The History of Norwegian Defence, Vol. 5, The Changing Character of Alliance Defence 1970–2000*] (Bergen: Eide forlag, 2004), pp. 39–41; Rolf Tamnes,

from the US, French and British concepts, which emphasised protection for submarines through covert patrols in the depths of the Atlantic.

The bastion concept developed gradually. Strategic submarines continued to patrol off the coasts of the US, and some of them also deployed to the southern hemisphere. By the late 1980s, however, most operated from bastions.[5] The bastion concept led to major changes in the Soviet Navy's strategic and operational priorities. A more distinct Soviet anti-access strategy emerged: the main mission of the Northern Fleet's general-purpose capability was now to protect and ensure the survival of the strategic submarines and their supporting infrastructure. To accomplish this mission, the fleet sought control of the Norwegian Sea, possibly down to between the Lofoten archipelago in northern Norway and eastern Greenland, and sea-denial down to the GIUK gap. This took priority over the traditional anti-SLOC mission: in a major war, relatively few attack submarines would probably be available for operations in the North Atlantic to sink US ships bound for Europe. However, they could still compel NATO to allocate forces to protect the SLOC.[6]

Consequently, Soviet naval and air activity in northern waters rose incrementally and exercises increased significantly, peaking with *Summerex* in 1985. The expansion undoubtedly made the northern flank a more prominent theatre of the Cold War.

NATO Response
The Soviet military build-up created a number of challenges for NATO. First, because of nuclear deterrence, NATO saw limited war as the likeliest type of conflict: in the words of the then US Secretary of Defense Robert McNamara,

The United States and the Cold War in the High North (Oslo: Ad Notam, 1991), pp. 28–29, 192–94.
[5] Håvard Klevberg, *'Request Tango', 333 skvadron på ubåtjakt – maritime luftoperasjoner i norsk sikkerhetspolitikk* ['*Request Tango', the 333 Squadron in Pursuit of Submarines – Maritime Air Operations in Norwegian Security Policy*] (Oslo: Norwegian University Press, 2012), pp. 345–50; National Museum of American History, 'Fast Attacks and Boomers: Submarines in the Cold War', <http://americanhistory.si.edu/subs/work/missions/warfare/index.html>, accessed 19 October 2016; Pavel Podvig (ed.), *Russian Strategic Nuclear Forces* (Boston, MA: MIT Press, 2000).
[6] Christopher A Ford and David A Rosenberg, 'The Naval Intelligence Underpinnings of Reagan's Maritime Strategy', *Journal of Strategic Studies* (Vol. 28, No. 2, April 2005), pp. 379–409; Vladimir Kuzin and Sergei Chernyavskij, 'Russian Reactions to Reagan's "Maritime Strategy"', *Journal of Strategic Studies* (Vol. 28, No. 2, April 2005), pp. 429–39; John B Hattendorf and Peter M Swartz (eds), *U.S. Naval Strategy in the 1980s. Selected Documents*, Newport Papers No. 33, (Newport, RI: Naval War College, 2008), p. 61.

a 'deliberate, surprise non-nuclear attack with limited objectives, e.g., an attempted "land grab" against Thrace, Hamburg, or Northern Norway'.[7] Second, the submarine threat underscored the need for defensive barriers, or chains of underwater listening posts, and more patrols of the Norwegian Sea.[8] As the Soviet naval build-up continued from the late 1960s and throughout the 1970s, NATO member states voiced doubts about their command of sea – the bridge between the US and Europe – and the US ability to fight forward and defend Allies. After a visit to Norway in 1971, the US Chief of Naval Operations Admiral Elmo R Zumwalt wrote: 'Norway feels increasingly behind the Soviet line as the result of her knowledge that NATO defense initially must be across Greenland/Iceland/UK Gap and because of the very high order of recent Soviet fleet exercises off Northern Norway'.[9] Other NATO members also had reason to ask if the US would risk the loss of carrier battle groups in order to attain sea-control in forward, high-threat areas. In the late 1970s, the Supreme Allied Commander Atlantic (SACLANT) assumed that a considerable number of Allied aircraft would have to operate from Norway to compensate for the non-deployment of carriers in the Norwegian Sea.[10]

These concerns did not at first elicit a Western maritime response. The initial NATO rebalancing in Europe and the north in the 1970s was air- and ground-centric: it prioritised the allocation of more land and air reinforcements. Plans for the defence of NATO's northern region, including the maritime domain, came to rely heavily on bases in the northern triangle: in northern Norway, Iceland and in the southern part of Northern Europe.

In the case of Norway, two initiatives were notable. First, in 1974 Norway was included in the US Air Force Co-located Operating Bases programme. The arrangement was gradually broadened to incorporate several air bases in northern Norway, reflecting a growing concern over the Soviet threat in the far north. Second, based on a bilateral agreement signed in 1981, the US Marine Corps assigned an amphibious brigade to the defence of Norway, and combined this with the pre-positioning of the brigade's heavy equipment. Britain, Canada and Germany also increased their commitments to flank defence during this period. In 1982, most of

---

[7] Tamnes, *The United States and the Cold War in the High North*, p. 200.

[8] Olav Njølstad and Olav Wicken, *Kunnskap som våpen. Forsvarets forskningsinstitutt 1946–1975 [Knowledge as a Weapon: The Norwegian Research Defence Establishment, 1946–1975]* (Oslo: Tano Aschehoug, 1997), pp. 212, 395–408.

[9] Tamnes, *The United States and the Cold War in the High North*, p. 235.

[10] *Ibid.* pp. 252–60.

the measures taken were incorporated into the Rapid Reinforcement Plan of the Supreme Allied Commander Europe (SACEUR).

The second phase – the forward maritime strategy and the accompanying US plan for a 600-ship navy – heralded a more powerful and self-confident Western maritime response to the Soviet maritime challenge. The strategy had its roots in a fundamental US re-examination of naval threats and counter-strategies of the late 1970s, maturing in the early 1980s and reaching its formative phase in 1984–86.[11] NATO's new maritime operations concept of 1981 embodied key elements of this new US maritime strategy. Unlike NATO's concept, the US maritime strategy had a global reach and focused increasingly on targeting Soviet strategic submarines in their bastions. Critics voiced concerns over nuclear stability.

Forward operations from the sea now constituted the US and NATO maritime strategy's pivotal contribution to a potential war with the Soviet Union. By putting pressure on Soviet bastions and the flanks, the USSR would be forced to use most of its naval forces to defend its strategic submarines and their supporting infrastructure, preventing Soviet Russia from deploying general-purpose forces further forward into the North Atlantic. In the event of a global naval war, therefore, it was unlikely that the Soviet Fleet would be able to focus its attention on the SLOC.

While the US and NATO maritime strategies differed in some respects, they both prioritised the Norwegian Sea. A conspicuous aspect of their plans was to deploy carrier groups to the north early in a crisis or war and place some of them in Norwegian coastal waters in Vestfjorden, north of Bodø. The concept was fully tested for the first time in 1985 during the NATO Exercise *Ocean Safari.*

The launching of the forward maritime strategy coincided with an equally important shift in the Western warfighting doctrine in Central Europe. In the 1970s, the build-up of Soviet bloc forces raised doubts about the credibility of a NATO defence that focused on defensive warfare and reinforcements. To deal with this problem, the AirLand Battle doctrine of 1982 heralded a more offensive orientation: land forces should be concentrated close to the border and used to win the first large-scale battles, while airpower should stop Soviet reserves from reaching the front.[12] With such a 'come as you are' doctrine, early reinforcements from the US would play only a minor role. This corresponded to key

---

[11] John B Hattendorf, *The Evolution of the U.S. Navy's Maritime Strategy, 1977–1986,* Newport Papers No. 19 (Newport, RI: Naval War College, 2004); Ford and Rosenberg, 'The Naval Intelligence Underpinnings of Reagan's Maritime Strategy', pp. 379–409.

[12] Douglas W Skinner, *Airland Battle Doctrine,* Professional Paper No. 463 (Alexandria, VA: Center for Naval Analyses, September 1988).

characteristics of contemporary maritime strategy, which also concentrated on forward operations and less on the transatlantic sea lanes and reinforcements. The Cold War therefore ended with more confident, offensive-oriented Western strategies in both the land and maritime domains, and both strategies differed significantly from thinking and planning in the 1960s and 1970s.

Actors and Interests: The Atlantic Command and Norway

Many actors took part in the formulation and implementation of NATO's strategy for the North Atlantic during the Cold War. European navies were given the task of containing a Soviet attack until large numbers of US forces could arrive, conducting sea-control operations in their own waters, and they added much to anti-submarine, amphibious and surveillance operations.[13] But the bulk of forces would come from the US, with the UK as an important second partner. Norway had a special role to play in the triangle because of its geostrategic location in the north and its tailored capabilities. This section presents a closer look at the role of SACLANT and Norway's policy and contributions.

In 1951–52, NATO set up two strategic commands: a European command led by SACEUR; and an Atlantic command led by SACLANT, based in Norfolk, Virginia. For most of the Cold War, SACLANT also held the position of Commander-in-Chief US Atlantic Fleet. His wartime tasks were sea-denial, to safeguard the sea lanes for reinforcements and resupply to Europe, and to maintain sufficient forces to project power. The basic structure of the command remained the same until it was dissolved in 2003. SACLANT had a number of important sub-commanders, notably Commander-in-Chief Eastern Atlantic, a British admiral with headquarters in Northwood, and Commander Striking Fleet Atlantic (STRIKEFLT), an American who was SACLANT's major subordinate seagoing commander.

As part of the American planning and decision-making structure, SACLANT provided the key institutional and operational bridge between the US and Europe. SACLANT had a deep understanding of the Soviet challenge from the north and in the maritime domain – in contrast to the continental focus of SACEUR. From the early 1980s, the forward maritime strategy rested predominantly on the deployment of STRIKEFLT to the Norwegian Sea. A major share of NATO's air reinforcements, and all of NATO's naval and amphibious reinforcements to northern Norway, were SACLANT forces. STRIKEFLT's presence off the Norwegian coast doubled the number of air-defence fighters and tripled the numbers of fighter-

---

[13] Geoffrey Till, 'Holding the Bridge in Troubled Times: The Cold War and the Navies of Europe', *Journal of Strategic Studies* (Vol. 28, No. 2, 2005), pp. 309–37.

bombers available to the Commander, Defence Command North Norway, who was also a principal subordinate commander under SACEUR. To support STRIKEFLT, a number of bases and installations for logistic support were built along the Norwegian coast.[14]

The bonds between SACLANT and Norwegian authorities were very close, especially since both focused increasingly on the north. Because Norway is a small country with modest resources, in a major war its own armed forces could hold on alone for only a short period until effective assistance arrived from abroad. The West German rearmament that began in the mid-1950s gave more substance to the forward defence of Scandinavia in the south and made it possible for Norway to concentrate more of its own forces in its north. Regardless, Norway calculated that these forces could not delay a Russian assault more than a few weeks. Hence a key objective of Norway's security and defence policy was to ensure the engagement of the Alliance. Over the 1980s, the size and quality of support from SACEUR and SACLANT improved significantly.

Committing the Alliance to the defence of the north was part of a calculus that also factored in Norway's border with the USSR. Norway thus saw the need to strike a balance between deterring and reassuring its far larger neighbour. The most pronounced restraint was in Norway's policy on bases, which precluded the permanent stationing of Allied combat forces in peacetime; an analogous ban against peacetime storing of nuclear ammunition; and curbs on Allied military activity in the very far north, particularly in Finnmark county. Critics argued that the Norwegian restraints weakened the defence of the country. However, Oslo contended that the dual policy was a well-designed balancing act, conceived in combination with credible plans for reinforcing Norway at short notice.

The dual policy was also expressed in the idea that drawing a firm line with the USSR could be combined with cooperation. Norway considered Soviet Russia's orientation as interest-based, so collaboration would be possible when deemed useful to both parties. This posture was influenced by history: Russia's behaviour in the north, unlike in other parts of Europe, never involved conquest and oppression. This part of the dual policy became very noticeable from the mid-1970s, when the Cold War intensified while Norway and the Soviet Union initiated closer cooperation on fisheries management.

This may give the impression that Norway was a net 'importer' of security. While this is true, Norway did not confine its efforts to defending its own territory; it also contributed to US and NATO grand strategy. For

---

[14] Jacob Børresen, 'Alliance Naval Strategies and Norway in the Final Years of the Cold War', *Naval War College Review* (Vol. 64, No. 2, 2011), pp. 97–115.

much of the post-1945 period, Norwegian shipping owners controlled 8–10 per cent of the world's merchant marine, and this fleet was a key asset in NATO's plans for ferrying supplies to the European theatre. Norway's most significant contributions were in early warning and intelligence collection. Due to its geographic proximity to the Kola region and the northern rim, Norway was in an advantageous position to chart the Soviet bomber force preparing for transpolar operations; to monitor air bases and air defences in the Leningrad–Kola area; to monitor vessels, submarines, pens and facilities as well as the deployment of naval forces; and to monitor nuclear test explosions and missile testing. From the early 1960s onwards, military surveillance satellites led to a revolution in intelligence, but did not reduce the need for permanent watch over the Soviet activities in the north.

This included close cooperation between the Norwegian Intelligence Service and US services. The collaboration included communications intelligence stations and installations, formalised in the NORUSA agreement of 1954, as well as a number of electronic intelligence and telemetry systems, brought together under one agreement, NORUSA II, in 1979.[15]

Another core activity of the Norwegian Intelligence Service, carried out together with foreign and domestic partners, was monitoring Soviet submarines. Underwater defence barriers were an important part of containing the Soviet Navy. The US built the Sound Surveillance System across the GIUK gap in the 1950s; later, Norway in collaboration with the US established several defensive barriers in the Norwegian and Barents Seas.[16] The 1980s saw many new initiatives, prompted by the more vigorous maritime strategy and by the increasing difficulty of detecting Soviet submarines. These initiatives included a new chain of underwater listening posts, vessels for acoustic research and intelligence collection and an extensive programme of military oceanographic, acoustic and bathymetric exploration from the Kara Sea to the Greenland Sea.[17]

Maritime patrol aircraft strengthened NATO against the Soviet fleet. Norway's own reconnaissance capability was very modest until the P-3 Orion entered service in 1969, which the US provided on favourable terms as part of bilateral surveillance and intelligence arrangements. Norwegian aircraft could now cover the northern part of the Norwegian Sea and the Barents Sea, while other NATO members could concentrate

---

[15] Olav Riste, *The Norwegian Intelligence Service, 1945–1970* (London: Frank Cass, 1999).

[16] Njølstad and Wicken, *Kunnskap som våpen*, pp. 212, 395–408.

[17] Rolf Tamnes, *Norsk Utenrikspolitikks Historie, bd. 6, Oljealder 1965–1995*, [*The History of Norwegian Foreign Policy, Vol. 6, Oil Age, 1965–1995*] (Oslo: Norwegian University Press, 1997), p. 74.

their attention further to the west and south. This geographic division of labour was embodied in an agreement in 1969 concerning Exercise *Gin Clear*, which also involved UK patrol operations from Scotland and US flights from Iceland. Leaving forward flights to Norway helped to reduce tension in the far north.[18]

From the Cold War to the 'New Normal'

The collapse of the USSR and deterioration of Russian military power removed concerns over the conventional military and naval threat. The Northern Fleet quickly fell into a state of crisis and decay. The number of operational and battle-ready submarines and surface vessels dramatically decreased. Attempts to build new major weapons systems, such as aircraft carriers and strategic submarines, encountered colossal problems. Some within the Russian military sector tried to sustain the bastion defence concept, primarily by concentrating most of the strategic submarines in the European Arctic.[19] However, after 1988, the Russian navy could not avoid a steady reduction in strategic patrolling, until the turn of the century when there were no boats on patrol. The northern region was marginalised in the new global geopolitics. In the north, Norway and Russia entered a period of closer cooperation based on common interests and a sense of shared purpose.[20]

The demise of the Soviet empire and the rising threat of international terrorism after 2000 had fundamental consequences for US and NATO security priorities and force structures. Most NATO countries reduced the size of their navies significantly, though far less dramatically than Russia, and prioritised quality and modernisation. Most importantly, the US and NATO could again enjoy command of the sea. This gave the naval forces a unique opportunity to project power on land in wars such as those in Afghanistan and Iraq; to secure good order at sea; and to devote their maritime resources to problems such as piracy, terrorism and weapons proliferation. Western naval forces – both those of Allies and partners – could do more to protect and sustain the interconnected global system. In the US, these ideas were embodied in *A Cooperative Strategy for 21st Century Seapower* of 2007, a document of the US Navy, Marine Corps and

---

[18] Klevberg, *'Request Tango'*, pp. 139–54, 267–80, 294–98.

[19] Kristian Atland, 'The Introduction, Adoption and Implementation of Russia's "Northern Strategic Bastion" Concept, 1992–1999', *Journal of Slavic Military Studies* (Vol. 20, No. 4, 2007).

[20] Rolf Tamnes and Kristine Offerdal (eds), *Geopolitics and Security in the Arctic. Regional Dynamics in a Global World* (London and New York, NY: Routledge Global Security Studies, 2014), pp. 167–77.

Coast Guard. In the words of Geoffrey Till, it had a strong flavour of optimism and post-modernism.[21]

NATO activity followed much of the same logic. From the 1990s, the main tasks were to support operations in the Balkans, Afghanistan and then Libya, and to protect law and order at sea. The traditional task of preparing for conventional war against a strong adversary disappeared entirely when NATO disbanded its Atlantic Command in 2003 and concentrated all its operational functionality in the hands of SACEUR. This coincided with the decision to abandon geographic commands. The new ones would be more cost-effective and flexible, but at the cost of area-specific knowledge and experience. All this meant that the Alliance no longer had a comprehensive strategy for the use of maritime forces. The new Alliance Maritime Strategy of 2011 saw a cautious re-examination, but its ambitions fell short of meeting emerging requirements.[22]

Geostrategic Shifts and Russia's Imperial Nationalism

The North Atlantic has reverted to an arena of potential conflict; but this development is not isolated from wider strategic shifts. Most dramatic is the apparent decline in the significance of geography due to the proliferation of weapons of mass destruction, international terrorism, the rising cyber threat to governments and the proliferation of high-precision, long-range strike weapons.

By the 1970s, Soviet theorists had anticipated that 'automated reconnaissance-and-strike complexes' would bring about a new military-technical revolution. The changes occurred first in the US as part of the historic shift associated with the Information Age. Until recently, the US has enjoyed a near-monopoly in precision-guided weaponry and the associated battle networks. Today, the proliferation of information networks and precision-strike capabilities is gathering momentum, allowing states such as China and Russia to extend the anti-access zones

---

[21] Chief of Naval Operations, Commandant of the Marine Corps and Commandant of the Coast Guard, *A Cooperative Strategy for 21ˢᵗ Century Seapower* (Washington, DC: Department of the Navy, Office of the Chief of Naval Operations, 2007); Amund Nørstrud Lundesgaard, 'Controlling the Sea and Projecting Power: US Naval Strategy and Force Structure after the Cold War', dissertation, University of Oslo, Faculty of Humanities, 2016, Chapter 6; Peter D Haynes, 'American Naval Thinking in the Post-Cold War Era: The U.S. Navy and the Emergence of a Maritime Strategy, 1989–2007', dissertation, Naval Postgraduate School, June 2013.

[22] Jo G Gade and Paal Sigurd Hilde, 'NATO and the Maritime Domain', in Bekkevold and Till (eds), *International Order at Sea*, pp. 116–39.

and threaten distant targets without necessarily projecting traditional military force.[23]

But geography remains significant, and three centres of gravity merit specific attention. The first is the rising importance of the Asia-Pacific region. The redistribution of global power will have great influence on the design of the international system and on US priorities. In the long run, the US emphasis on the Asia-Pacific in response to a rising China makes it harder to maintain a strong footprint in Europe, which calls for a more equitable sharing of the transatlantic burden.[24]

The second centre of gravity stretches from West Africa to Pakistan. This presents profound social, economic and political problems, including terrorism, migration and regional war – all with implications for the defence of NATO's southern flank.

The third centre of gravity is Europe, particularly because under Putin, Russia's most important foreign policy goal is to restore essential parts of the empire and establish buffer zones along its borders. Russian wars in Georgia and Ukraine have demonstrated its ability to use military force, pressure and intimidation to achieve political goals. A second important goal for Putin is to remake Russia as a twenty-first century great power, demonstrated by its intervention in Syria.

Will Putin succeed? The future does not belong to Russia. After nearly a decade of growth, Russia's economy and political system have descended into Soviet-era stagnation. Reactionary policies go hand-in-hand with imperialist nationalism. At the heart of the Kremlin's view of world affairs and muscle-flexing lies the traditional and instinctive Russian sense of insecurity – a centuries-old advance of uneasy nationalism that has always blurred the distinction between defensive and offensive actions.[25] In the long run, the combination of an over-ambitious, authoritarian and dysfunctional regime, economic decline and potential domestic unrest may lead the country into a new national tragedy.[26] States in decline are often unpredictable and occasionally aggressive.

---

[23] Andrew F Krepinevich, 'Maritime Competition in a Mature Precision-Strike Regime', Center for Strategic and Budgetary Assessments, 2014.

[24] Aaron L Friedberg, *Beyond Air-Sea Battle: The Debate over US Military Strategy in Asia* (London: International Institute for Strategic Studies, 2014).

[25] *The Economist*, 'The Fog of Wars', 22 October 2016; John Lewis Gaddis, *George F. Kennan: An American Life* (New York, NY: Penguin Press, 2011), p. 220 – referring to Kennan's 'long telegram' of 22 February 1946; and Fiona Hill and Clifford G Gaddy, *Mr. Putin: Operative in the Kremlin* (Washington, DC: Brookings Institution Press, 2013).

[26] Kjell Grandhagen, 'Russia, the Arctic and the Ever-Changing Security Environment', *Three Swords Magazine* (Vol. 29, 2015), pp. 44–49.

In the foreseeable future, economic hardships and structural problems in the military–industrial complex will slow down Russia's current military modernisation effort.[27] This may induce the leadership to seek accommodation with the West in certain areas. However, a new direction in foreign and security policy is not likely. The world must live with an assertive and militarily strong Russia.

NATO member states are collectively stronger than Russia, should they muster their combined economic and military resources. Yet they have not done so. Russia's relative strength is not purely numerical. First, it is in a favourable geostrategic position to influence and dominate significant parts of its former empire. Second, unlike NATO, the Russian leadership is less bound by formal institutions and thus can be quicker in decision-making and operations. Third, Russia is well prepared to pursue hybrid warfare and operations in grey zones, exploiting its adversary's inability to coordinate. Subversion, disinformation and forgery, combined with the use of special forces, constituted the heart of Soviet-style operations and have reappeared under Putin. By using such methods, Russia may hope to reach its goals before the West can respond with conventional means. Fourth, Russian armed forces are about to acquire an offensive-oriented structure, with capabilities tailored for large-scale war and an ability to quickly deploy well-trained forces.[28]

Russia's strategy of denying NATO access to land and sea is visible across Europe. The Black Sea Fleet in particular gives Russia the potential to project power and block NATO access throughout Southeastern Europe. In the course of a few years, Russia has also regained a position of strength in the Baltic Sea, making it difficult for NATO to reinforce Allies and partners. In both regions, Russia's strategy is based on the deployment of multiple short- and medium-range missile systems, particularly the S-400 Triumf, Iskander and Kalibr missiles.[29]

Russia's objectives in the far north are no less ambitious than its aims further south. The Arctic is of great economic and military strategic importance for Russia and helps to maintain the country's stature as a great power. Climate change and melting ice will likely open new Arctic waterways, including passages to Asia, and increase the ability to exploit

---

[27] For an in-depth analysis, see Gudrun Persson (ed.), *Russian Military Capability in a Ten-Year Perspective – 2016* (Stockholm: FOI-R-4326-SE, 2016), pp. 133–48, 189.

[28] See, *inter alia*, Igor Sutyagin, 'Russia Confronts NATO: Confidence-Destruction Measures', RUSI Briefing Paper, July 2016.

[29] Stephan Frühling and Guillaume Lasconjarias, 'NATO, A2/AD and the Kaliningrad Challenge', *Survival* (Vol. 58, No. 2, April–May 2016), pp. 95–116; and Jonathan Altman, 'Russian A2/AD in the Eastern Mediterranean: A Growing Risk', *Naval War College Review* (Vol. 69, No. 1, Winter 2016), pp. 72–84.

the region's huge mineral and petroleum resources. While this economic potential of the far north exerts a strong influence on Russian thinking and planning, much of it cannot be realised until the distant future.

Russia's current agenda in the far north centres on traditional security threats – notably incoming ballistic and cruise missiles – and the need to address societal security challenges, such as ship accidents, oil spills, sabotage and smuggling. Russia is about to rebuild and modernise many of its air bases and to deploy intermediate- and long-range air-defence systems on the continental portion of the Arctic rim, from Murmansk to Chukotka, and on many island territories, including Franz Josef Land, Novaya Zemlya, the New Siberian Islands, Wrangel Island and Schmidt Island. Russia is also spending significant economic resources to protect the Northern Sea Route and on building stations for search-and-rescue operations.[30]

Equally importantly, forward bases in the north enable the deployment, dispersal and support of the bombers normally stationed further inland. In 2007, Russia resumed regular strategic flights with Tu-95MS Bear H and Tu-160 Blackjack bombers over the Arctic and along European coasts. The sorties are often combined with shorter flights by non-strategic bombers such as the Tu-22M3 Backfire and Su-24 Fencer. In addition to testing NATO's air defence and wartime contingencies, Russia also uses the flights for political posturing.

Last, but not least, while Russia displays its strategy of denying NATO access throughout Europe, the Northern Fleet and the bastion defence stand out as a *strategic* challenge to transatlantic defence by threatening the link between North America and Europe. Russia has re-established the bastion strategy, reaching a stable level of activity from 2008. As during the Cold War, the patrol area is primarily in the western bastion in the Arctic and in the eastern bastion in the Pacific Ocean.[31]

The bastion defence concept remains essentially the same as during the Cold War. Defensive and offensive actions are intertwined and indistinguishable. In a conflict, Russia will seek to protect its strategic forces, which would involve establishing sea-control in its immediate vicinity and sea-denial further west and south, down to the GIUK gap. Some attack

---

[30] See, *inter alia,* Atle Staalesen, 'New Arctic Military Bases Soon Manned and Equipped', *Independent Barents Observer,* 11 July 2016; *Independent Barents Observer,* 'New Russian Military Facilities Planned and Under Construction', 1 November 2016.

[31] A Russian military spokesman stated in 2015 that Russia had also returned to the practice of deploying nuclear ballistic missile submarines to the southern hemisphere. *RT,* 'Russian Nuke Subs to Patrol Southern Seas for First Time since Soviet Era', 1 June 2015.

Figure 1: Reach of the Bastion Defence.

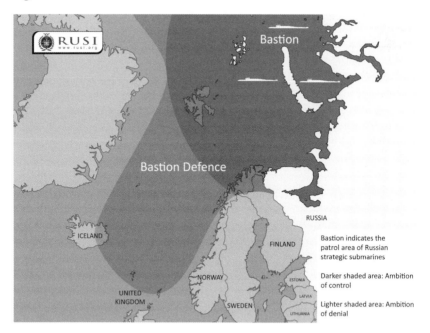

submarines will most likely also operate further west in the Atlantic. Such operations would weaken NATO's ability to project forces into Europe.[32]

Russia's confrontational mode of operations – not only in the Baltic region but also in the North Atlantic – exacerbates NATO's concerns. Russian attack submarines have been observed off the east coast of the US and outside Faslane in Scotland, the home port of the UK's nuclear submarines. The complex energy and transport infrastructure has become vulnerable to maritime hybrid warfare, as detailed in Chapter VI. Russia operates underwater vehicles and surveillance vessels that can cut vital undersea cables carrying commercial and military data between the US and Europe.[33]

---

[32] Expert Commission on Norwegian Security and Defence Policy, *Unified Effort – Expert Commission on Norwegian Security and Defence Policy* (Oslo: Norwegian Government Security and Service Organisation, 2015), pp. 20–23.

[33] Kathleen H Hicks et al., *Undersea Warfare in Northern Europe* (Washington, DC: Center for Strategic and International Studies, 2016); *IISS Strategic Comments*, 'Russia's Naval Modernisation Will Take Time' (Vol. 21, Nos 7–8, November 2015); and Thomas Nilsen, 'Shipyard Reveals Unique Video of Spy Submarine', *Independent Barents Observer*, 12 November 2016.

Figure 2: Missile Systems – Position and Range.

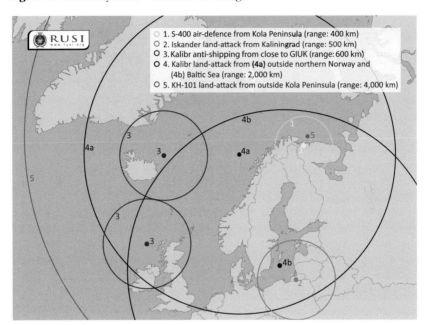

To strengthen planning and operations in the Arctic and North Atlantic down to the GIUK gap, Russia established the Arctic Joint Strategic Command in 2014, directly under the General Staff and with the Northern Fleet as its main striking force. This command has taken control of various forces that were previously part of the Western, Central and Southern Military Districts. The command and the fleet have become the pivot in Russia's anti-access strategy in the maritime domain.[34]

Strategic submarines constitute the core of the Northern Fleet. At present, the fleet has seven. The *Delta IV* class was delivered in 1984–90 and will remain in service until around 2030. Constructing the next generation posed immense difficulties for many years, but the Russians succeeded in the end. The first boat of the *Dolgorukiy* class, equipped with the Bulava missile, entered service in 2013, and others are currently being brought into service with the Northern and Pacific Fleets. Russia has contracted for eight submarines altogether.

---

[34] *GlobalSecurity.org*, 'Arctic Strategic Command Sever (North) Unified Strategic Command (USC)', <http://www.globalsecurity.org/military/world/russia/vo-northern.htm>, accessed 4 November 2016.

Table 1: Russian Navy Strategic and General-Purpose Forces 2015.

	Northern Fleet	Baltic Fleet	Black Sea Fleet	Caspian Flotilla	Pacific Fleet	Total
SSBN	7				5	12
SSGN/SSN	17				9	26
SS	6	2	4		8	20
Submarines						**58**
CV	1					1
CGN	2					2
CG	1		1		1	3
DD	4				4	8
DDG	1	2	1		2	6
FFG		1				1
FF		6	2	2		10
Major surface vessels						**31**
FFL	6		6		9	21
PGG	6	11	9	4	15	45
PG		7		4		11
LST	4	4	7		4	19
LCU		2		1		3
Minor surface vessels						**99**

Source: Office of Naval Intelligence, The Russian Navy: A Historic Transition *(Washington, DC: Office of Naval Intelligence, 2015).*

Russia currently has 23 attack submarines in service in the Northern Fleet.[35] These boats have been crucial to protecting the strategic submarines in the bastions and in sustaining Russia's effort to restrict NATO's freedom of movement. Today, about half of the submarine fleet is ready to put to sea at any given time.

The fleet has been reduced significantly, both in terms of absolute numbers since the Cold War and relative to the US fleet. At the end of the Cold War, Russia had more than twice as many submarines as the US; today, the fleet is around 80 per cent the size of the US force. That said, the reductions are less dramatic than in many other sectors of the Russian armed forces, and modernisation of the fleet has high priority. That includes the existing *Oscar II-* and *Akula*-class submarines and first and

---

[35] For a study of capabilities and potential, see Igor Sutyagin, 'Russia's New Maritime Doctrine: Attacking NATO's Sea Lanes of Communication in the Atlantic – Intent and Feasibility', *RUSI Defence Systems* (21 August 2015); and Igor Sutyagin, 'Russia's New Maritime Doctrine: Attacking NATO's Sea Lanes of Communication in the Atlantic – Sustainability, Future Capabilities and Potential Countermeasures', *RUSI Defence Systems* (28 August 2015).

foremost the new multipurpose *Severodvinsk*, which became combat-ready in early 2016 after many years of delay due to financial problems – considered to be the least detectable Russian submarine, albeit not as quiet as some of the US classes. The *Severodvinsk* can carry both conventional and nuclear cruise missiles and launch strikes against land targets, submarines and ships. In the context of the Norwegian Sea, of significance is that the *Severodvinsk* is designed to attack surface naval group formations with long-range anti-ship missiles. Russia plans for the procurement of seven boats.

In addition to the large and expensive *Severodvinsk*, Russia is designing a smaller multipurpose submarine, Project *Husky*-class, which will probably appear in different versions: one with anti-submarine armament; and another with long-range anti-ship armament. Using this new class of boat to protect strategic submarines will allow the multi-purpose *Severodvinsk* to perform other missions.

The modernisation of the surface fleet is also demanding. Russia has introduced new and capable smaller ships, particularly corvettes and frigates, but failed to build or complete new major warships that make up the foundation of the blue-water capacity. The last capital warship to be completed was the battlecruiser *Pyotr Velikiy* (*Peter the Great*), which entered service in 1998. Russia's only aircraft carrier, the *Admiral Kuznetsov*, dates from 1995 and has been plagued by many problems. Russia plans to build a new-generation carrier, but the fate of the project is uncertain.

Russia has invested heavily in precision-guided missile technology that will have a profound impact on its deterrence and defence capability, especially since it will increasingly enable Russia to threaten distant targets without deploying a traditional power-projection naval or air force. The growing armoury of missiles is important to sustain bastion defence and to protect strategic submarines, but their role is much broader than sea-denial; they are central to defending Russia in general and in denying the US access to Europe.

Four types of missile systems illustrate Russia's anti-access strategy.[36] First, air defence remains a high priority. Russia has based S-400 ground-launched anti-aircraft missiles near the Norwegian and Northern European borders. The range of the missiles (250–400 km) constitutes a significant challenge to defending northern Norway. The problem will be further

---

[36] Ørjan Askvik, '*Utviklingen av langtrekkende konvensjonelle presisjonsvåpen – konsekvenser for Norges evne til avskrekking og forsvar mot angrep*' ['The Development of Long-Range Conventional Precision Weapons – Consequences for Norway's Ability to Deter and Defend Against Attack'], master's thesis, Oslo, Norwegian Defence University College, Spring 2015.

aggravated by the introduction of systems such as the S-500 (55R6M), which has the accuracy, speed and range to intercept and destroy intercontinental ballistic missiles as well as hypersonic cruise missiles and aircraft.[37]

Second, Iskander tactical missiles constitute one of the most important weapons in Russia's anti-access strategy. They are accurate, with a range of 400–500 km. They can be armed with either conventional or nuclear warheads and appear in both ballistic and cruise variants. They are deployed in Kaliningrad and other parts of the Western Military District; in the Baltic region, Russia has used Iskander deployments as a political weapon against NATO. A new system, the R-500 long-range cruise missile has been tested and will bolster Russia's effort to build a solid buffer zone.

Third, the Kalibr missile – a modular system with anti-shipping, land-attack and anti-submarine variants – has an operational range of up to 2,500 km. It can hold distant targets at risk using conventional or nuclear warheads.[38] Russia plans to deploy Kalibr capability on all newly designed submarines, corvettes, frigates and larger surface ships.

Fourth, there is Russia's growing ability to project force by using air-launched long-range cruise-missiles. The conventionally armed Kh-101 probably has a range of up to 4,000 km. This provides Russia with the ability to strike high-priority targets with precision from long range. The stealth capability, high-subsonic speed and low-altitude flight profile of these missiles make defence against such weapons very difficult. These examples illustrate the strength and breadth of Russia's anti-access strategy.

NATO Renewal

How should Western states respond to Russia's military build-up and muscle-flexing, particularly in the maritime domain? NATO has become a heterogeneous organisation, but remains by far the best option that exists. The Alliance has been slow to respond to a resurgent Russia. However, the Wales Summit in 2014 and the Warsaw Summit in 2016 demonstrated that NATO, despite internal cleavages, is still able to decide on a strategy and follow up with concrete military measures. Strong US leadership has been vital to the Alliance: NATO cannot exist without the US. But adequate US engagement in Europe, embodied in the NATO alliance, cannot be taken for granted in the future – even less so should the Europeans not shoulder a larger share of the burden.

---

[37] See, *inter alia*, Dave Majumdar, 'Russia's Deadly S-500 Air-Defense System: Ready for War at 660,000 Feet', *National Interest,* 3 May 2016.

[38] See Office of Naval Intelligence, 'The Russian Navy: A Historic Transition' December 2015, p. 34.

In the past two years, NATO's response to an increasingly assertive Russia has been air- and ground-oriented. It has concentrated much of its attention on building forward presence in the Baltic States and Poland, relying mostly on land and air forces. Military geography favours Russia. One school of thought contends that NATO must build a strong, direct defence to counter this advantage. According to one study, preventing the rapid overrun of the Baltic States alone would require about seven brigades, adequately supported by airpower and land-based fires and ready to fight at the onset of hostilities.[39] Another school of thought asserts that the Western states should have confidence in NATO defence and deterrence. The crux of the strategy is to win the battle of the mind: to convince the opponent that the cost of an assault is higher than the benefit and that the Alliance *will* escalate if necessary. This is in fact NATO's implicit strategy today: the Alliance is not willing to muster a direct defence in forward areas strong enough to halt Russian assaults, so the answer is to establish tripwires in the centre and in the north.[40]

Reassuring NATO members and deterring Russia has so far involved three measures concentrated on the eastern flank. The first is the establishment of a persistent presence of four multinational Allied battalions on a rotational basis in the three Baltic States and Poland, each consisting of some 800–1,000 troops. The second measure is an extension of the US presence on the eastern flank through the rotational deployment of an armoured brigade combat team, headquartered in Poland, and the pre-positioning of materiel. The third consists of the redesign of the NATO Response Force (NRF), in particular through the formation of the Very High Readiness Joint Task Force (VJTF). This force of about 5,000 soldiers, supported by air and sea combat assets and special forces, can deploy within days: it is designed to convey the message to an adversary that an assault would trigger NATO's defence guarantees and a more powerful Western response. While developing this concept, NRF follow-on forces will also be made more robust and available for operations.

However, NATO still lacks a coherent approach to its southern and northern flanks. NATO is updating its geographic defence plans – including a Graduated Defence Plan for Norway, Iceland and the maritime flanks – which will create a foundation for concrete measures. But questions remain. What is the prospect of Allied engagement and support in the north? Can the VJTF contribute or is there significant risk that it will

---

[39]   David A Shlapak and Michael W Johnson, 'Reinforcing Deterrence on NATO's Eastern Flank: Wargaming the Defense of the Baltics', RAND Corporation, 2016.
[40]   See, *inter alia*, Alessandro Scheffler Corvaja, 'Beyond Deterrence: NATO's Agenda after Warsaw', Konrad Adenauer Stiftung: Facts & Findings – Prospects for German Foreign Policy, No. 224, October 2016.

be tied down in other places, especially on the eastern flank? Can the task force reach a battlefield in the far north at short notice if an assault comes with little or no warning? The VJTF will be better tailored for a northern crisis scenario when the entire joint force – with its air, maritime and special forces components and strategic airlift – is in place, and even more so if modules of the force could be singled out and arrive without delay to become part of a tripwire.

These reflections underscore the need to think beyond existing NATO forces and to develop other multilateral and bilateral arrangements. The US Marine Corps continues to be relevant in Northern Europe, as it was during the latter part of the Cold War, based on the bilateral agreement of 1981. Equipment and materiel for a Marine Task Force of about 4,500 marines are pre-positioned in mid-Norway. From January 2017, the corps has also placed a rotational force of 300–400 marines at the Værnes Air Station in mid-Norway, giving a boost to training and exercises in Northern Europe and making the US guarantee more credible. The UK's broad assortment of capabilities can also play a potentially crucial role in northern crisis scenarios: a carrier strike force to support amphibious capability; maritime patrol aircraft once reintroduced; and the Joint Expeditionary Force. In November 2016, Britain and Norway reaffirmed their intent to increase defence cooperation, both by working more closely together on maritime surveillance and by signing a new agreement on host-nation support, further increasing their ability to exercise, train and operate together.

Eventually, NATO planning should view the north as one single theatre of operations, incorporating not only Alliance members and the maritime domain, but also the non-aligned states of Finland and Sweden. Their cooperation with NATO has become so comprehensive that it resembles a quasi-alliance, forming a functional defence community, but one not protected by NATO's Article V guarantee. Should Finland and Sweden join NATO, Russia would react strongly, but their membership in the Alliance would strengthen the defence of Northern Europe significantly.

The Alliance is going through a process of renewal, but the strength and pace of the effort are insufficient. The tasks that lie ahead specifically in securing the maritime domain are no less daunting. As discussed in depth in the following chapters, NATO will have to take a number of steps to confront Russia's anti-access strategy. Six considerations merit special attention:

1. Since Russia is on the verge of establishing long-range anti-access capabilities that would reach all of Europe, any maritime strategy for the North Atlantic must include both the Baltic region and the Norwegian Sea.

2. The defence of NATO's northern flank and maritime domain would benefit greatly from integrating UK and Norwegian flights with US operations out of Iceland.
3. A reformed command structure should reintroduce geographical areas of responsibility, a strong command for the maritime domain with close ties to US headquarters, and tight connections between NATO headquarters and national and multinational headquarters.
4. Improved situational awareness and information sharing are vital to maintain knowledge about Russia's intentions and capabilities. The UK, Norway and the US are about to expand their surveillance capabilities for operations in the maritime domain and in the northern region.
5. Forces and bases must be made less vulnerable to precision-guided strikes, cyber and hybrid attacks, and a premium should be put on survivable forces that can operate within contested zones. More forces must be made available for the battle of the seas and littorals, both by reinvigorating standing naval forces and by introducing next-phase response forces.
6. More extensive training and Article V exercises are imperative, both to communicate cohesion and strength and to prepare for combat operations in high-threat scenarios. Norway will host the high-visibility Exercise *Trident Juncture* in 2018, which could represent an important step towards a new maritime strategy.

The Norwegian Contribution

Norway is currently readjusting its armed forces away from out-of-area operations towards a new posture focused on Russia. What role can Norway play in the defence of the north and the battle of the seas?

Three types of new strategic capabilities have the highest priority: purchasing 52 F-35s to replace the F-16 fleet; acquiring four new submarines from Germany to replace the *Ula* class; and enhancing the quality of intelligence and surveillance, including the acquisition of five P-8 Poseidon maritime patrol aircraft to replace the P-3 fleet.[41] These investments supplement five very capable *Fridtjof Nansen*-class frigates received in 2005–11. Apart from renewing the force structure, Norway deems it necessary to improve the sustainability and reduce the vulnerabilities of the defence organisation.

Norway's new defence measures aim to better prepare it to discover hostile actions and respond without delay. This would contribute to the

---

[41] Norwegian Ministry of Defence, *Capable and Sustainable: Long-term Defence Plan* (Oslo: Norwegian Ministry of Defence, 2016).

rebuilding of a credible forward presence. However, Norway still depends heavily on military assistance from abroad and must prepare for assaults with little or no warning. To a greater extent than before, credible deterrence and capable defence require both the *simultaneous* involvement of Allies and *seamless* escalation – initially to establish an Allied tripwire, and subsequently to bring in stronger forces with more firepower.[42] This assessment should guide NATO defence planning for the north and the maritime flanks.

Although the rise of the new Russia under Putin has led Oslo to concentrate on the defence of the north, Norway's contributions stretch beyond defending its own territory. The military participates in international operations, especially in the Baltic and in the Middle East, and Norwegian capabilities are a significant contribution to NATO's endeavour to counter Russian anti-access efforts and secure the maritime domain.

Intelligence collaboration lies at the heart of the broader Norwegian–American military partnership, and this cooperation is about to expand in scope and quality. Reconnaissance and anti-submarine warfare capabilities are essential to sustain a robust NATO maritime strategy. The contours of a collaborative club of P-8 countries is emerging with UK aircraft operating from Scotland, US planes from Iceland and Norwegian ones from northern Norway. The division of labour makes political, economic and operational sense, but also contributes to maintaining low levels of tension in the north.

This leads back to Norway's dual-track policy. While balancing and deterring Russia is an imperative, Norway and Russia must inevitably work together in the north. The collaboration does not suffer from the strains seen in the Baltic region, which are created by Russia's aggressive conduct and deep mutual distrust. International treaties, institutions and regimes help to shape the northern region in important areas such as fisheries, environment, and search and rescue. The Norwegian armed forces contribute to this through cooperation with the Russian authorities on matters pertaining to the coast guard, border guard and search-and-rescue operations. In view of Russia's military build-up in a strategically important and sensitive region, frequent contact between the two states contributes to stability as it reduces the risk that signals will be misunderstood, while also reducing the risk of unintentional provocation and inadvertent escalation.

---

[42] Commission on Norwegian Security and Defence Policy, *Unified Effort*, pp. 61, 66–68.

Conclusion

Russia has returned to its historical mode: its fear of encirclement goes hand in hand with aggression and infiltration abroad. The outside world may have to live with an assertive and militarily strong Russia for a long time. A key part of Russia's strategy is to deny NATO access to land and sea areas around the country. The Northern Fleet and the bastion defence concept present a *strategic* challenge to the link between North America and Europe.

While burden-sharing remains an issue, NATO has demonstrated that it is able and willing to respond to Russia's military build-up. The Alliance has so far concentrated on establishing forward presence in the Baltic States and Poland, based mostly on land and air forces. But NATO still lacks a coherent approach to its northern flank and the maritime domain. Because Russia's anti-access strategy could affect all of Europe, NATO maritime strategy must stretch beyond the North Atlantic to include the Baltic and Norwegian Seas.

From its front line position, Norway has a role to play in watching Russia, in contributing to the credibility of NATO's deterrence and defence – and in keeping the door open to Russia. After all, security is a combination of military strength and political dialogue.

II. THE UK AND THE NORTH ATLANTIC AFTER BREXIT

MALCOLM CHALMERS

This chapter analyses the UK's role in the North Atlantic in the context of historical and contemporary debates on how to balance across a range of defence commitments, preserving flexibility while optimising for priority threats, at a time when the European security order is at its most unsettled since the end of the Cold War.

The Atlantic in UK Strategic Priorities

The UK's position as the leading economic and military power in the mid-nineteenth century was closely related to its possession of the world's most powerful navy, providing security for global trade and a far-flung empire. As this hegemonic position began to be eroded from the late nineteenth century onwards, however, the government was increasingly forced to make difficult choices, giving up influence in some areas – notably in the Americas under the Monroe Doctrine – in order to focus on others – for example, protecting imperial sea routes to India and the Far East. The trade-off was further sharpened in two successive world wars, with the surrender of Singapore in February 1942 symbolising the UK's inability to successfully fight a two-front war.

The emergence and consolidation of the wartime alliance with Washington allowed the UK to partially restore a global role to the Royal Navy immediately after 1945, albeit one that was clearly subordinate to that of the US. As decolonisation gathered pace, however, the strategic case for prioritising Britain's military presence 'east of Suez' diminished under the dual pressures of budgetary austerity and NATO demands for more credible conventional capabilities in Europe. In 1966, the government announced the cancellation of the programme to build a new class of aircraft carrier, arguing that it no longer justified a high priority in a more NATO-centric defence strategy. This was followed, in 1968, by the cancellation of the order for F-111K long-range strike aircraft for the same reason.

For the next two decades, the two primary lines of maritime effort were, first, maintaining and guarding the nuclear-powered ballistic missile submarines (SSBN) force, which took over the country's strategic deterrent role from 1968; and, second, making a key contribution to the Alliance's anti-submarine warfare capabilities in the Northeast Atlantic. NATO's sea lines of communication would be crucial in the event of a prolonged war in Central Europe. The UK's naval and air forces, together with Allied forces based in the UK, would have played a key role in protecting those reinforcements.

This focus on nuclear and NATO roles was never absolute, with the Royal Navy successfully maintaining a range of globally deployable capabilities. Less than a year after the 1981 Nott Review, which had consolidated the focus on nuclear and anti-submarine warfare capabilities at the expense of amphibious and light carrier capabilities, the 1982 Falklands War provided a lifeline for those who believed in the continuing value of expeditionary forces. Over the following decades, the requirement to be able to defend the Falklands against a future attack was to provide a compelling rationale for large-scale investments in long-range maritime power.

The sudden collapse of Soviet power in Europe after 1989 had an even more profound effect on the balance of British military effort. Significant cuts in UK commitments to NATO's front lines in Central Europe and the North Atlantic followed, allowing an overall reduction in real defence spending of some 19 per cent between 1990/91 and 1998/99.[1] The post-Cold War period also saw a marked revival in the UK's focus on expeditionary operations. More than any other European power (with the partial exception of France), the UK has maintained both an interest in, and capabilities for, conducting military intervention at distance. The first example of this focus came in 1991, when the UK deployed more than 50,000 troops as part of the US-led coalition to liberate Kuwait from Iraq.

This pattern – of the UK providing by far the largest European contribution to US-led coalition operations – was to repeat itself several times over the next two decades. Even as the role of possible conflict with Russia as a force driver for conventional capabilities continued to decline, British ambitions for the use of military power in 'out-of-area' operations grew. The government committed significant forces to a series of

---

[1] For further discussion, see Malcolm Chalmers, 'The 2015 SDSR in Context: From Boom to Bust – and Back Again?', *RUSI Journal* (Vol. 161, No. 1, February/March 2016), p. 5.

increasingly ambitious operations in Bosnia, Kosovo, Afghanistan and Iraq, in each case determined that the UK could and should play a leading role in transforming those societies for the better. Between the 'force for good' 1998 Strategic Defence Review and the 2010 Strategic Defence and Security Review (SDSR), total defence spending rose by 30 per cent in real terms.[2]

This period was to prove to be a high point of British globalism. Under the dual pressures of the 2008 global economic recession and the perceived failure of military intervention in Iraq and Afghanistan, the UK's appetite for a global military role faltered and its defence budget fell. Between 2009 and 2016, its NATO-declared defence budget (including spending on operations) fell from 2.5 per cent to 2.2 per cent of GDP.[3] Spending in real terms dropped by some 17 per cent between 2010/11 and 2015/16.[4]

The importance of national security in the formulation of defence and security priorities also began to increase after 2010. This reflected both a degree of disillusionment with force for good interventions (intensified after the 2011 Libya operation) and growing concern over direct threats to the UK, especially from terrorism and cyber attack. An increased focus on national security became a central element in both the 2010 and, especially, 2015 SDSRs. It has gathered pace with the appointment of Theresa May as prime minister in July 2016: as home secretary over the previous six years, May had been focused primarily on countering direct threats to the UK and its citizens. This shift of focus towards a policy driven more by national interest is likely to be further deepened by the repatriation of many of the civilian instruments of foreign policy from the EU to UK national control over the next three years.

Maritime Programmes and Choices

Maritime capabilities did not play a central role in any of the major military interventions of the past quarter of a century, with the partial exception of Sierra Leone. Yet the Royal Navy has not been neglected in intra-Ministry of Defence resource allocation. Programmes to buy advanced air-defence

---

[2] Between 1998/99 and 2010/11. *Ibid.*

[3] Facing similar pressures, defence spending by the US fell even more sharply: from 5.3 per cent in 2009 to 3.6 per cent in 2016. On a comparable measure, only five countries – Estonia, Latvia, Lithuania, Poland and Romania – increased the proportion of their GDP spent on defence during this period, with Norway maintaining its spending at 1.54 per cent of its GDP. See NATO, 'Defence Spending of NATO Countries (2009–2016)', PR/CP(2016)116, press release, 4 July 2016, Table 3.

[4] Malcolm Chalmers, 'Spending Matters: Defence and Security Budgets after the 2015 Spending Review', RUSI Briefing Paper, May 2016, Table 1.

destroyers (the Type 45) and *Astute*-class nuclear attack submarines, both with global range, are nearing completion. Two large aircraft carriers, equipped with F-35B aircraft, are due to enter service by the end of the decade, creating a fleet carrier capability that has not existed since the 1970s. Work on building the new *Dreadnought* class of SSBNs is also ramping up, in what is now estimated to be a £31-billion capital programme. As a result of these combined commitments, spending on strategic nuclear and naval systems is set to take around half of the total equipment budget for the next decade.[5]

In order to maintain backing for this ambitious investment programme, however, the Royal Navy was forced to accept sharp reductions in personnel, together with a steady reduction in the numbers of frigates and destroyers. The 2015 SDSR has promised some respite from these trends. The process for building the eight new anti-submarine Type 26 Global Combat Ships promised in the SDSR is well advanced, with steel set to be cut on the first ship in summer 2017. The SDSR also committed the government to buy nine new P-8 maritime patrol aircraft off the shelf, with entry into service from 2019. Taken together with its highly capable fleet of SSN attack submarines, the UK is therefore on track to have three potent (albeit numerically limited) capabilities for anti-submarine warfare in the North Atlantic, should it choose to deploy them in this role. At the same time, all three platforms are globally deployable, able to be used flexibly in response to future developments in the strategic environment.

A further significant injection of funds into maritime budgets, over and above those already committed, is unlikely before 2020. While the UK has reaffirmed its commitment to continue to spend at least 2 per cent of its GDP on defence, this has translated into an increase of only 1 per cent in real defence spending between 2015/16 and 2019/20. The depreciation of sterling in the aftermath of the Brexit vote, if sustained, could add a further £700 million to annual spending on the procurement of imported military systems, mainly from the US. The SDSR's commitment to developing the ability to deploy a national expeditionary force of some 50,000 by 2025 (of which between 30,000 and 40,000 would be from the army) adds further pressure on a limited defence budget.

The Russian Strategic Shock

When making difficult choices on how to meet Treasury targets for spending cuts, the 2010 SDSR prioritised the need to maintain a credible

---

[5] Ministry of Defence, *The Defence Equipment Plan 2016* (London: The Stationery Office, 2015), p. 17.

counterinsurgency capability in Afghanistan, where the UK still had more than 10,000 troops fighting a difficult campaign in Helmand province. In contrast, the review gave relatively little attention to the role of future conflict with major powers as a possible force driver.

The Russian annexation of Crimea in 2014, and its subsequent intervention in eastern Ukraine, changed this thinking. By the time the next SDSR was published in November 2015, the government had concluded that defence planning must now take much more account of the possibility of future state-based threats, with particular reference to Russia. As part of this reappraisal, the 2015 review devoted an extended section to the importance of deterrence, a concept largely underplayed during the previous decades in which the focus had been on the defeat of non-state actors or weak states. In doing so, the government emphasised:

> We will use the full spectrum of our capabilities – armed force including, ultimately, our nuclear deterrent, diplomacy, law enforcement, economic policy, offensive cyber, and covert means – to deter adversaries and to deny them opportunities to attack us.[6]

As a consequence, a process of recalibration of defence priorities has begun. The relevance of so-called 'legacy' capabilities – supposedly relevant only to the deterrence of other militarily capable states – is now being reasserted. With the completion of its withdrawal from Helmand at the end of 2014, the army is beginning to work through the implications of reorienting itself towards the very different challenges of confronting an opponent that can deploy sophisticated air and ground forces, denying any easy haven to a British expeditionary force. Air-defence interceptors – of limited value in Iraq and Afghanistan – have now become a critical element of UK capability, tracking the growing number of Russian aircraft incursions into the skies around the British Isles, while also contributing to NATO's enhanced air policing mission in support of the Baltic republics. Not least, the rationale for the Trident-armed nuclear deterrent force, irrelevant in responding to non-state opponents in Iraq or Afghanistan, has become clearer in the context of contributing to the deterrence of a well-resourced, nuclear-armed power such as Russia.

Moreover, as time has passed since the start of the Ukraine crisis, opinion has hardened on the permanence of the shift in Russian policy towards a more confrontational and adventurist stance. The large and sustained Russian intervention in Syria since late 2015, together with its subsequent unwillingness to agree to a ceasefire as it pressed for a rebel

---

[6] HM Government, *National Security Strategy and Strategic Defence and Security Review 2015: A Secure and Prosperous United Kingdom*, Cm 9161 (London: The Stationery Office, 2015), p. 24.

surrender in Aleppo, led to a further hardening of views within Whitehall during 2016. Even if the Ukraine crisis was the trigger for the deterioration of NATO–Russia relations over the past three years, the confrontation now appears to have become broader and more intractable.

As a consequence, the UK government believes, NATO needs to make clear that it is prepared to invest in credible deterrent capabilities that ensure that Russia does not miscalculate the commitment of the Alliance – including the UK – to respond appropriately, and decisively, to any aggression against a member state.

The first steps towards this goal were taken at the September 2014 Wales Summit. The spending pledges made there are now bearing fruit, with most of NATO's European members increasing their defence budgets in real terms between 2014 and 2016. Five of the most exposed Eastern European states – Poland, Estonia, Latvia, Lithuania and Romania – have already reached the 2 per cent target (in the cases of Poland and Estonia) or will soon do so. After many years of gradual decline, total NATO European spending on defence has risen in real terms in both 2015 and 2016.

This turnaround has also been evident in the UK. Despite initial (and embarrassing) hesitancy in the immediate aftermath of the Wales Summit, the government is now committed to maintaining spending at a minimum of 2 per cent of GDP through to 2020. The UK has also played a leading role in collective military preparations, both as part of NATO and as part of coalition counter-Daesh operations in Iraq and Syria. In October 2016, the UK announced the deployment of some 800 personnel to Estonia, where it will lead NATO's enhanced forward presence in that country.

The decision to purchase a new generation of US-made maritime patrol aircraft, taken in the last stages of the 2015 SDSR process, also reflected the growing importance that deterrence of Russia has taken as a driver of capability choices. Because of the expense involved, and the other competing calls on the procurement budget, maritime patrol capability became one of the hardest fought elements of the SDSR. While the entry into service of this new capability appears to have been somewhat delayed because of budgetary constraints, the decision to forgo domestic procurement options is a powerful indication of the government's acceptance of the military rationale for this new capability. It reflects both a strong belief that this is a capability that is now needed to counter new risks that Russia could pose in coming decades, and a belief that it would not be wise, or credible, to rely on Allies to fill this gap.

In contrast to the more radical 'east of Suez' supplementary defence reviews of 1967 and 1968, the 2015 SDSR does not reflect a fundamental shift of strategic focus away from a global orientation and towards a primary focus on Europe. Perhaps the strongest thematic element of the review was the need to focus more attention on non-conventional threats such as cyber attacks, terrorism and subversion. The authors of the review were aware that, even in relation to Russia, the risks to NATO in coming years are likely to combine traditional military threats with an increased focus on this wider range of instruments, which Russia might think can help it to achieve its aims while avoiding a direct confrontation with the Alliance's superior aggregate conventional capabilities.

Brexit and Trump

The strategic shocks have been arriving with increasing frequency since 2014. Russian aggression in Ukraine was followed by the dramatic success of Daesh (also known as the Islamic State of Iraq and Syria, or ISIS) in Iraq, the intensification of the Syrian Civil War and an upsurge in terrorist attacks across both the Middle East and Europe. By 2015, largely as a result of intensifying conflicts across the region, migrants fleeing across the Mediterranean and Aegean Seas for Europe had reached unprecedented levels. The attempted coup in Turkey in July 2016, and the clampdown that followed, has intensified the rift between one of NATO's most important members and other member states.

Yet, when this period is examined in retrospect, it could be political developments within Western democracies that will turn out to have been most consequential for UK and European security.

The first such shock began with Prime Minister David Cameron's unexpected victory in the May 2015 general election meant that he had no alternative but to honour his promise to hold a referendum on the UK's membership of the EU. The consequences of the clear victory in June 2016 for the Leave campaign that followed are not yet clear. Once the UK has left the EU, it will no longer be represented in the main European institutions that develop and agree common foreign policies. It will also need to devote considerable political and administrative energy to establishing national control over a wide range of policy instruments currently organised at EU level, including trade policy, agriculture, fisheries, regulation of state aid, sanctions, financial regulation, climate change and (some part of) development aid. This combination of reduced influence on collective deliberations along with new national responsibilities is likely – at least in the short and medium term – to deepen the already apparent trend towards a more inward-looking UK approach to the problems of the wider world. While the UK's contribution to collective military capabilities through NATO will be largely unaffected, its ability to shape wider foreign and security policy discussions could be undermined.

Second, the UK is only beginning to come to terms with the election of Donald Trump as US president, promising a radical shift in policy towards an 'America First' approach. At the time of writing in January 2017, it still seems probable that the US will remain strongly committed to maintain its security guarantees to European NATO members, even as it continues to press for greater financial contributions from those (such as Norway) which spend less than 2 per cent of their GDP on defence.

A third shock – Scottish independence – has failed to materialise, with the people of Scotland voting by a margin of 55 per cent to 45 per cent against separation in the September 2014 referendum. Yet the prospect of a future referendum remains a potent factor in the debate on what Brexit will mean, as does the future of Northern Ireland. It is no longer possible to understand the future politics of the North Atlantic without also having some appreciation of the multinational character of the UK state.

Both the UK's decision to leave the EU and Trump's election are testament to the power of nationalism as a political force in two of the West's hitherto most stable democracies. They could yet prove to be only the start of a broader trend, now spreading across Western Europe. Even after it leaves, the UK's future – as with other non-members – will continue to be shaped by the political and economic fate of the EU. There will be some who argue that the UK's departure should allow it to bear a lesser share of the burden of protecting the EU's eastern and southern frontiers. They may suggest that the UK should leave it to others – such as Germany and France – to take more of the load in protecting an EU that professes to become more of an autonomous strategic actor.

Yet an opposite argument can be made in relation to the UK's immediate neighbourhood in Northern Europe and the North Atlantic. As Peter Hudson and Peter Roberts argue in Chapter V in this Whitehall Paper, the UK is probably the only European power that could credibly take command of NATO maritime forces in the North Atlantic, given its comparative advantage in the relevant capabilities. For a UK outside the EU, this might be a useful element in a broader (albeit not rigid) division of labour between the major European powers. As long as the US provides the bulk of the capabilities available for any particular crisis, the level and disposition of European forces – including in the North Atlantic – may matter somewhat less. If the US were to reduce its commitment to European security, however, the need for a strong UK contribution to the continent would increase, potentially even at the expense of its capabilities for global power projection.

The fate of the UK's economy will continue to be a central determinant of its appetite for further defence investments. Its commitment to an ambitious programme of liberal internationalism during the Blair and Brown years – reflected in growing aid as well as defence budgets – was rooted in a period of rapid economic growth (averaging 3.3 per cent annually in real terms between 1991 and 2007) without parallel in the post-war period. But if the pessimists are right, then the social and political strains within British society are likely to become more evident as the economic costs of Brexit become ever clearer. The appetite for the UK to play an ambitious international role could decline further, with clear national needs becoming the main driver for any new defence and security commitments. But an optimistic scenario is also possible. Viscount Matthew Ridley, for example, has suggested that 'there's little doubt that we could find ourselves growing at 4 or 5 per cent a year' if the UK were to take full advantage of the 'golden opportunity' for free trade that Brexit could shortly provide.[7] If this were to be the case, then it would be possible to devote substantially more resources to the defence budget after 2020, while still allowing more money to be spent on health, social care, infrastructure and other pressing national needs.

Where is Russia Going?

Since the 2014 annexation of Crimea, it has become conventional wisdom in NATO circles that the new Russian threat is here to stay, and that this will continue to be the central security risk facing Europe for the foreseeable future.

While NATO must prepare for this possibility, however, a simple extrapolation of recent trends could prove to be an unreliable predictor of future events. As a consequence of the sharp reduction in its oil and gas revenues since 2014, alongside the costs of economic sanctions imposed as a result of its Ukraine actions, Russia is now entering a period of deep economic crisis. Total GDP is broadly similar to that of South Korea or Australia, and around half that of the UK or France.[8] Living standards are falling and deep cuts are being made in government spending on health and education. Having doubled since 2008, the defence budget is now due to fall by some 15 per cent in real terms between 2015 and 2019.[9] The official Russian defence budget for 2017 is 2.835 trillion roubles, equivalent to only $47.9 billion as of late January. Even after due allowance is made for other items not included in the official figure, total

---

[7] Matt Ridley, 'A Golden Opportunity for a Free-Trade Bonanza', *The Times*, 3 October 2016.

[8] International Monetary Fund, 'World Economic Outlook Database', October 2016, <https://www.imf.org/external/pubs/ft/weo/2016/02/weodata/index.aspx>, accessed 25 January 2017.

[9] Kathrin Hille, 'Russia Prepares for Deep Budget Cuts that May Even Hit Defence', *Financial Times*, 30 October 2016.

Russian defence spending probably amounts to less than a tenth of NATO spending, which totalled some $920 billion in 2016.[10]

In these circumstances, some commentators are suggesting that powerful elites may be pushing for an easing of tensions with the West as a means of escaping from the economic cul-de-sac in which Russia finds itself.[11] It is still possible that President Vladimir Putin and his most security-minded advisers will push ahead with a security-first approach, perhaps even to the extent of reintroducing further elements of a command economy (for example, placing new limits on capital exports and even expropriating privately owned assets). If further steps in this direction were to be considered, however, they could face strong resistance from oligarchic interests. Outside analysts simply cannot predict what the outcome of such conflicts would be.

The election of President Trump introduces a further element of uncertainty into the picture. The new US president might be seen by some in the Russian leadership as providing an opportunity for further adventurism, for example in relation to NATO's exposed Baltic member states. Yet it might also provide an opportunity for another attempt at a reset of NATO–Russian relations, based on mutual concessions on issues of concern.

While the broader course of relations between NATO and Russia may be less certain than many think, this very unpredictability is an added reason for NATO to maintain a deterrent posture that is credible, measured and adaptable. Given the disparity in overall defence efforts between the Alliance and Russia, a full-scale NATO mobilisation for a new Cold War is neither necessary nor stabilising. Rather, NATO members need to be prepared both to respond proportionately to further Russian aggressive behaviour, and to be ready to match Russian confidence-building measures with constructive responses of their own.

Conclusions

The state of European security has not been as unsettled as it is today since the end of the Cold War. In addition to pressing security concerns to its east and south, European liberal democracies now face severe challenges from nationalist opponents who seek, more or less coherently, the dismantlement of the post-1945 institutional architecture and its replacement with a 'Europe of nations'. While the Brexit vote in the UK

---

[10] NATO, 'Defence Spending of NATO Countries (2009–2016)'.
[11] Nick Butler, 'Can the Oligarchs Save Russia?', *Financial Times*, 17 October 2016; Adam Taylor, 'A Russian Academic Suggested that Putin Might Step Down in 2017. Then His Prediction Disappeared', *Washington Post*, 12 November 2016.

followed by Trump's election are so far the most visible elements of this backlash, they may not be the last.

In this uncertain world, NATO continues to provide a strong element of stability, limiting the degree of renationalisation in defence and reducing the chances of miscalculation by an unpredictable Russian leadership. The UK remains a firm supporter of NATO and has demonstrated since the Brexit vote that it remains committed to playing a leading role in generating the capabilities that it requires.

Faced with the political and economic consequences of Brexit, there is a risk that the UK could move further towards a more inward-looking approach to foreign and security policy, focused more on the protection of its direct security interests and less on the country's role as a global security provider. Even if that were to be the case (and it may not), the defence of the North Atlantic will continue to be seen as an essential element in the defence of the UK's most vital interests.

III. THE CENTRALITY OF THE NORTH ATLANTIC TO NATO AND US STRATEGIC INTERESTS

JOHN J HAMRE AND HEATHER A CONLEY

On 4 April 1949, twelve countries assembled in Washington, DC to sign the Washington Treaty. At the onset of the Cold War, the North Atlantic represented both a geostrategic imperative and a unifying ideal of collective defence shaped by the need to protect its members from a growing Soviet threat. Sixty-eight years and 29 NATO members later (once members ratify the membership of Montenegro), it is time for NATO to return to and relearn its founding principles in a twenty-first century context.

However, 2017 is not 1949, even though a major power is once again threatening the international order. An increasing number of state and non-state adversaries threaten the security and prosperity of NATO members, while a rise in transnational threats and significant technological disruption also negatively impact the security of NATO members.

Perhaps more challenging is the waning of societal strength, credibility and assurance in the transatlantic community. Public opinion in NATO countries is neither fully knowledgeable about, nor supportive of, NATO's purpose. A 2015 poll suggests that the majority of citizens in Germany, France and Italy believe their country should not use military force to defend a NATO Ally involved in a serious military conflict with Russia.[1] As historical memories of the Cold War fade, the Kremlin will continue to exploit perceived democratic weaknesses – exposed by

---

[1] Danielle Cuddington, 'Support for NATO is Widespread among Member Nations', Factank: News in the Numbers, Pew Research Center, 6 July 2016.

disenfranchised populations which reject globalisation and supranational bodies in favour of nationalist and protectionist policies – through the deployment of active measures and *kompromat* (the collection of compromising material for blackmail) in conjunction with twenty-first century cyber offensive tools.

Since Russia's military intervention in Georgia in 2008, its illegal annexation of Crimea and intervention in eastern Ukraine in 2014 and military operations in Syria in 2015 and 2016, NATO members have slowly awakened to the need to provide a credible deterrence and defence posture in Europe. The new situation has some Cold War-like elements, but lacks the stabilising rules of conduct of the Cold War era. The established international rules and norms – ranging from maritime law to the non-proliferation regime, the use of armed drones and hybrid tactics in armed conflict as well as terrorism – which the two superpowers developed during and after the Cold War are no longer adequate.[2]

For these reasons, there must be a clarion call for a new US and NATO security approach to the North Atlantic region. NATO, and in particular the US, must relearn the development and execution of its defence reinforcement plans in Europe to counter Russia's growing multilayered anti-access and power projection capabilities. More specifically, NATO must also understand that the North Atlantic – and in particular the Arctic – is an increasingly important part of Russian military strategic calculation, as evidenced by its growing defence modernisation efforts as well as naval and air prowess.

It is therefore essential that the North Atlantic region comes to be seen as being central to NATO's own strategic interests and be a recipient of more NATO assets. This new approach must prioritise maritime, air and missile defence capabilities, substantially increase regional exercises and drive greater bilateral and regional defence efficiencies. NATO should assess whether to return to a modernised and flexible regional command structure specifically for this region. This should only be done, however, if such a command structure supports greater regional defence capabilities while minimising bureaucratic structures and redundancies. A more focused and capable NATO in the North Atlantic domain will enhance Alliance unity, afford NATO members enhanced security and protection from external threats and reaffirm international rules and norms – something that the signatories of the 1949 Washington Treaty would have recognised and understood.

---

[2] Richard Fontaine, 'How Trump Can Save the Liberal Order: And Even Reform and Strengthen It', *Foreign Affairs*, 30 November 2016.

A Distracted and Reactive US Power

NATO was built on the existing US-led allied structures from the Second World War and was galvanised by US strategic vision, a US investment plan to rebuild Europe and make it more resilient against communist ideology, and an enduring US military presence in Europe.

Although the existential threat to Europe dissipated with the collapse of the Soviet Union in 1991, this benign security environment began to change dramatically in 2007–08 as NATO expanded further towards Russia's borders. At the same time, many of the countries bordering Russia were experiencing political turmoil, with strong sections of the population rejecting Russia's influence over their economic, political and security policies. Economically strengthened by high energy prices and global financial developments, Russia sought to arrest NATO's influence by invading Georgia in August 2008. After the invasion, tensions between NATO and Russia temporarily flared and interactions were suspended but shortly thereafter, a new US president reset relations with Russia and a less antagonistic Georgian government was elected, which reduced tensions but essentially froze Russia's occupation of parts of Georgia. At the time of writing, Russian forces remain in Georgia and 20 per cent of its territory continues to be occupied.

The relative focus and clarity of US and NATO positions and policies during the Cold War have given way to today's ambivalence and hesitation. US and European defence assets, while large and technologically superior, are finite and are experiencing readiness and modernisation challenges as they stretch across three major theatres of operation. Transatlantically, there is limited strategic vision and political will to confront challenges to the international order.

Traditional tactics of warfare – disinformation and influence operations – have been deployed by Russia in new, highly effective twenty-first century technological forms. The use of offensive cyber and electronic warfare capabilities – ranging from crippling a power grid in Ukraine to hacking and revealing the contents of American emails to influence the outcome of the US presidential election – have sapped transatlantic political will.[3] Russia is thus able to exploit the weaknesses in the US and Europe while they confront a crisis of confidence in their own democratic and economic systems.

Russia seeks to instil fear and confusion within NATO member societies, aiming to destabilise the international system and upend legal

---

[3] Pavel Polityuk, 'Ukraine Investigates Suspected Cyber Attack on Kiev Power Grid', *Reuters*, 20 December 2016. Ellen Nakashima et al., 'Top U.S. Intelligence Official: Russia Meddled in Election by Hacking, Spreading of Propaganda', *Washington Post*, 5 January 2017.

norms that were once accepted by the Soviet Union and the Russian Federation. Annexation as seen in Crimea had not occurred in Europe since the Second World War. Russia has been in violation of the 1987 Intermediate-range Nuclear Forces Treaty for the past eight years. Senior Russian government officials have publicly raised the possibility that Russian tactical nuclear weapons could be deployed under certain circumstances and have repositioned nuclear-capable surface-to-air missiles in Kaliningrad. Russia has also suspended the 1990 Treaty on Conventional Armed Forces in Europe.

In sum, a lack of transatlantic resolve and Russia's military activities in support of its goal to return to great power status will create an environment where a weakened West will acknowledge Russia's status and accommodate Moscow's strategic interests, reversing the course of the past 25 years.

Russia's Strategic Calculations in the North Atlantic

The North Atlantic region today reflects the military muscle memory of Russia's past strength, its quest for present and future recognition of its great power status and its ability to prevent US forces from reinforcing their military capabilities in Europe. Strategically, Russia has reprioritised the region, as reflected in recent changes to its military and maritime doctrine. As during both the Second World War and the Cold War, Russia recognises and prioritises the North Atlantic as a critical sea line of communication (SLOC). This is where Russia can deploy its nuclear-capable forces from the Kola Peninsula and prevent US forces from deploying to reinforce NATO forces on its northern and eastern flanks. NATO's ability to maintain a credible deterrence rests on US capacity and other members to reinforce its force deployment should the likelihood of conflict between Russia and NATO increase. Thus, Russia has increased its exercise tempo and presence in the Arctic as well as in the North Atlantic, demonstrating capabilities to disrupt these sea lanes while purposefully increasing its forward operating presence, coastal defences and area denial capabilities concentrated in both the North Pacific and the North Atlantic.

First and foremost, Russia seeks to protect the freedom of navigation and access of its strategic nuclear deterrent based within its Northern Fleet on the Kola Peninsula. Russia has increasingly demonstrated its growing maritime capabilities in the region, particularly related to submarine (both nuclear and diesel) and surface ship capabilities which have long-range (conventional and non-conventional) missile capabilities. Russia has also demonstrated its ability to deploy advanced sea mines and underwater drones, which could hinder NATO and US access to these waters. Over the past several years, Russia has deployed nuclear-capable long-range bombers along the Eastern Atlantic coast and US Pacific coast to demonstrate range, power projection and ease of operation.

Since 2007, Russia has steadily expanded its military presence and infrastructure in the Arctic, including President Vladimir Putin's order to resume regular air patrols over the Arctic Ocean.[4] For example, Russian bombers penetrated the North American Aerospace Defense Command's (NORAD) 12-mile air defence identification zone around Alaska eighteen times during 2007.[5] In June 2014, NORAD scrambled two F-22s after four Russian long-range strategic bombers and an accompanying refuelling tanker entered the air defence identification zone, within 200 miles of Alaska.[6] That same month, Russian aircraft appeared to engage in a simulated attack on the heavily populated Danish island of Bornholm.[7] In August 2014, there were reports of multiple breaches of Finnish airspace by Russian state aircraft.[8] In September that year, two Russian military aircraft crossed into Swedish airspace south of the island of Öland.[9] The Russian navy, particularly the Northern Fleet, has been rapidly developed and modernised and military bases in the Russian Arctic have reopened. In addition, Russia has held large and complex military exercises, and substantially increased its military presence in the Arctic through the creation of Arctic brigades and command centres.[10]

A significant increase in major military exercises is also telling. For instance, in 2013, an exercise carried out around the Kola Peninsula demonstrated a more streamlined command structure and the deployment of more efficient tactical units. In March 2015, Putin announced a snap exercise in the Arctic that consisted of 45,000 Russian forces, fifteen submarines and 41 warships at full combat-readiness. Moscow has also announced plans for a total of fourteen operational airfields in the Russian Arctic, 50 airfields by 2020 and a 30 per cent increase in Russian special

---

[4] Ariel Cohen, 'Russia in the Arctic: Challenges to U.S. Energy and Geopolitics in the High North', in Stephen J Blank (ed.), *Russia in the Arctic* (Carlisle, PA: Strategic Studies Institute, 2011), p. 21.

[5] *Ibid.*

[6] Dan Whitcomb, 'NORAD Scrambled Fighters after Russian Bombers Seen Off California Coast', *Reuters*, 12 June 2014.

[7] David Blair, 'Russian Forces Practiced Invasion of Norway, Finland, Denmark and Sweden', *The Telegraph*, 26 June 2015.

[8] Kati Pohjanpalo and Kasper Viita, 'Finland's Fighter Jets on Alert as Russia Violates Airspace', *Bloomberg*, 29 August 2014.

[9] Thomas Frear et al., 'Dangerous Brinkmanship: Close Military Encounters Between Russia and the West in 2014', Policy Brief, European Leadership Network, November 2014.

[10] Heather A Conley and Caroline Rohloff, *The New Ice Curtain: Russia's Strategic Reach to the Arctic* (Lanham, MD: Rowman & Littlefield, 2015), p. 9.

forces based there, which suggests that the Arctic has emerged as a major theatre of operations for Russia.[11] Since 2014, Russia has conducted its largest and most complex military exercises which integrate air, land and sea components and simulate repelling an external attack by a very capable opponent.

Finally, growing military capabilities based in the Arctic and deployed in the North Atlantic are playing an increasing role in rebuilding Russia's great power status and fuelling its new form of nationalism and national identity. The Patriarch of the Russian Orthodox Church visited the home of Russia's Northern Fleet in August 2016, noting that 'the Russian Arctic has always played an important role in the fate of our motherland'. In 2015, Russian Deputy Prime Minister Dmitry Rogozin, on a visit to the North Pole, was accompanied by Russian orthodox priests who sprinkled holy water. Rogozin noted that 'the Arctic is Russia's Mecca'.[12] Another signal of Russian intentions in the Arctic is its growing investment in its readiness. In June, Russia launched the world's largest and most powerful icebreaker – the *Arktika*.[13] It is the next in line of a growing fleet of ice breakers which totals 40 active ships, with six more under construction and two more planned.[14]

Centrality of the North Atlantic to NATO Strategic Interests

It is for these reasons that NATO and the US must return to the North Atlantic in a more agile and technologically capable way. They must protect the vital SLOC and exercise reinforcement capabilities in a scaled-down version of the Cold War-era *Return of Forces to Germany* (*Reforger*) exercises, when the deployment of US and Allied forces to Europe via air and sea was tested. While there has been a dramatic uptick in NATO exercises since the Russian annexation of Crimea in 2014, these exercises have not focused specifically on reinforcement across the North Atlantic. While several exercises have been dedicated to the Baltic Sea region, they were not of the scale, scope and complexity of similar Russian exercises (45,000 forces in the North Atlantic in 2015 and 100,000 forces in the Russian Far East in 2014). In 2016, Norway hosted *Dynamic Mongoose 2016*, an

---

[11] Heather A Conley, 'Russian Strategy and Military Operations', statement before the US Senate Armed Services Committee, 8 October 2015.

[12] Ishaan Tharoor, 'The Arctic is Russia's Mecca, Says Top Moscow Official', *Washington Post*, 20 April 2015.

[13] Camila Domonoske, 'Russia Launches World's Biggest, Most Powerful Icebreaker', *NPR*, 1 June 2016.

[14] Jen Judson, 'The Icebreaker Gap', *Politico*, 1 September 2015.

annual NATO-led anti-submarine warfare exercise involving more than 5,000 troops from Alliance members and partners.[15]

In 2017, Norway will host *Trident Javelin*, a large joint force exercise that will test the Alliance's air capabilities and its ability to quickly secure air superiority.[16] The following year, it will host *Trident Juncture*, NATO's largest exercise.

Historically, NATO reinforcement exercises were based on the Alliance's ability to freely access and navigate the North Atlantic SLOC. Russia has prioritised modernising its submarine and long-range bomb capabilities to deny NATO unfettered access to the North Atlantic. Russia's enhanced submarine presence in the Greenland-Iceland-UK (GIUK) gap hinders US reinforcement strategies.

Regionally, in October 2016 Russia acknowledged that it has positioned the Iskander-M tactical nuclear ballistic surface-to-air missile system, which has a range of up to 450 miles, in its exclave of Kaliningrad on the Baltic Sea coast.[17] Home to Russia's Baltic Sea Fleet, the Kremlin's focus on Kaliningrad and the general military readiness of the region has led to an increase in the presence of Russian military vessels, specifically the *Buyan-M*-class corvettes, which provide considerable missile capability and range (Kalibr nuclear-capable cruise missiles which have a range of up to 930 miles). The Russian government has also enhanced its coastal defences in Kaliningrad with Bastion and Bal land-based anti-ship missile systems, which have a range of over 150 miles.[18] The range of the Pionersky Radar Station on the northern coast of Kaliningrad extends across Europe and can give an early warning of an incoming air attack.[19]

There is also uncertainty about command and leadership issues related to Russian forces in Kaliningrad. The Kremlin relieved the commander of the Baltic Sea Fleet and approximately 50 senior officers due to the fleet's poor state of readiness and reluctance to more

---

[15] NATO, 'NATO Launches Anti-Submarine Warfare Exercise in Norwegian Sea', 20 June 2016, <http://www.nato.int/cps/en/natohq/news_132596.htm?selectedLocale=en>, accessed 26 January 2017.

[16] Maro V Schanz, 'All for One in NATO', *Air Force Magazine* (Vol. 98, No. 10, October 2015), p. 30.

[17] *Deutsche Welle*, 'Russia Moves Nuclear Missiles onto NATO's Doorstep', 8 October 2016; Dmitry Solovyov and Andrius Syats, 'Russia Moves Nuclear-Capable Missiles into Kaliningrad', *Reuters*, 8 October 2016.

[18] Andrew Osborn and Simon Johnson, 'Russia Beefs up Baltic Fleet amid NATO Tensions: Report', *Reuters*, 26 October 2016.

[19] Lidia Kelly, 'Russia's Baltic Outpost Digs in for Standoff with NATO', *Reuters*, 5 July 2016.

aggressively engage US and NATO forces operating in the region.[20] Moscow has recently increased its military presence in Kaliningrad. Two small missile ships armed with Kalibr cruise missiles have already been spotted in the Baltic Sea and may soon join a newly formed division in Kaliningrad. According to Russian military sources, a further three small warships armed with the same missiles could join the fleet by 2020.[21]

Russia has repeatedly demonstrated that it wishes to return to Cold War-era military engagement where it was internationally recognised as a superpower – with Russian fighter jets flying within 10 feet of the USS *Donald Cook* in the Baltic Sea in April 2016.[22] While the US military sees such actions as provocative and unprofessional military behaviour, its Russian counterpart views this as 'normal practice' – and its absence over the past fifteen years as a lack of proper military exercising. There have been numerous Russian violations of the airspace of NATO members and partner countries by Russian fighter jets in the Baltic Sea region, the most recent being Russian Su-27 fighter jets infiltrating Finnish airspace on 6 October 2016.[23]

Past, Present and Future: US Force Posture in Europe

US security commitments in Northern Europe and the North Atlantic region have largely waned since the end of the Cold War, most notably in the reduced presence of land forces across the region. At its apex in the late 1980s, the US maintained approximately 340,000 permanently stationed military personnel in Europe to deter the conventional threat that the Soviet Union and Warsaw Pact forces posed to West Germany and Western Europe. In addition, the US maintained large stockpiles of pre-positioned equipment in Western and Northern Europe – enough for several divisions and support units – to allow forces based elsewhere to rapidly reinforce the continent in the event of a conflict. The US and NATO Allies annually rehearsed this reinforcement capability with the *Reforger* exercises, which by the late 1980s involved up to 100,000 US and Allied troops.[24]

---

[20] Elizabeth Schumacher, 'Kremlin Makes Sweeping Purge of Baltic Fleet Commanders', *Deutsche Welle*, 1 July 2016.
[21] Osborn and Johnson, 'Russia Beefs up Baltic Fleet amid NATO Tensions'.
[22] Barbara Starr et al., 'U.S. Issues Formal Protest to Russia over Baltic Sea Incident', *CNN*, 14 April 2016.
[23] Lisa Ferdinando, 'Russian Airspace Violations in Nordic-Baltic Region Dangerous, Work Says', *U.S. Department of Defense, DoD News, Defense Media Activity*, 7 October 2016.
[24] Kathleen H Hicks et al., 'Evaluating Future U.S. Army Force Posture in Europe', Center for Strategic and International Studies, June 2016.

As the Warsaw Pact collapsed in 1989 and the threat from the Soviet Union diminished, the US began a rapid drawdown of the US Army's presence in Europe. This process continued throughout the 1990s despite conflict in the Western Balkans, when the presence stabilised at approximately 60,000 soldiers.[25]

Amid a stable security environment in Europe and growing demand for US forces in Afghanistan and Iraq, the Bush administration announced in August 2004 the return of 70,000 US troops stationed overseas, of which 40,000 were to be removed from Europe over a six- to eight-year period. Thus, the heavy armoured brigades of the 1st Armored Division and 1st Infantry Division returned to the US and the US Army presence was further reduced to roughly 28,000 troops by 2012. Over the next several years, the heavy forces, enablers and headquarters elements of the 1st Armored Division and 1st Infantry Division were gradually reassigned from Germany to the US.

In late 2007, the then US Secretary of Defense Robert Gates temporarily halted the withdrawal of the last two heavy brigades from Europe due principally to a lack of basing for the troops in the US and the concerns of US military commanders in Europe that armoured capabilities were necessary to meet theatre security requirements. This left the US Army force posture in Europe at approximately 40,000 soldiers.[26]

In Northern Europe specifically, there were significant US force posture changes. In Norway, at the height of the Cold War, the US maintained an active duty military presence on land and at sea. From 1950 to 1990, active duty personnel ranged from 49 to 215, with a peak of 1,437 in 1976.[27] Beginning in 1982, Norway housed US pre-positioned military equipment in an elaborate cave system that has been undergoing a modernisation process since 2012. Today, these caves hold a variety of vehicles including armoured amphibious vehicles, high-mobility multipurpose wheeled vehicle variants, medium tactical vehicle replacements and logistics vehicle system replacements, snow-capable tracked vehicles, tank recovery vehicles and trailers, and towed carriages which are prepared for global US Marine Corps deployment.[28] At sea, Norwegian as well as Allied submarines used Olavsvern near Tromsø as a supply and service base for the Norwegian and Barents Seas. Olavsvern

---

[25] *Ibid.*
[26] *Ibid.*
[27] US Department of Defense, Defense Manpower Data Center, 'DoD Personnel, Workforce Reports & Publications', <https://www.dmdc.osd.mil/appj/dwp/dwp_reports.jsp>, accessed 27 January 2017.
[28] Christopher P Cavas, 'Cave-Dwellers: Inside the US Marine Corps Prepositioning Program-Norway', *Defense News*, 20 September 2015.

was also NATO's closest naval base to Russia's submarine bases along the coast of the Kola Peninsula, to the west of Murmansk.[29]

Important force posture changes were also made to the US defence presence in Iceland. Keflavik Naval Air Station was first used as an air bridge to Europe during the Second World War. It later deterred the advance of the Soviet Union, when the US military patrolled Icelandic airspace and conducted anti-submarine missions in the North Atlantic as part of the 1951 US–Iceland Defense Agreement.[30] However, by 2006, the US had withdrawn its four F-15s and several helicopters based at Keflavik, as well as more than 1,200 US military personnel, 100 Defense Department civilian employees and accompanying search-and-rescue assets.[31] In 2008, the US presence was replaced by a rotating NATO Icelandic air policing presence, which was also performed by non-NATO members Sweden and Finland. In response to an increase in Russian submarines patrolling the North Atlantic, it was announced in February 2016 that the US would rotate P-8 maritime patrol aircraft to Keflavik.[32] While the deployment of the P-8s revives a rotational US presence in Iceland, deployments will solely consist of maritime patrol aircraft for short durations. Eventually, the US Navy could establish regular patrol rotations.[33]

NATO, Missile Defence and the North Atlantic

It is also important to note NATO missile defence capabilities and US missile defence architecture in the North Atlantic and in Northern Europe more broadly. NATO adopted missile defence as a core Alliance mission in its 2010 Strategic Concept,[34] to which the US contributes. Although the European Phased Adaptive Approach (EPAA) is not designed to defend the GIUK region or the North Sea, the system represents a longstanding and major irritant to Russia, particularly due to its geographic location. The first two phases of EPAA have been completed to include an Aegis Ashore facility in Deveselu, Romania, which achieved initial operating capability status at the July 2016 Warsaw Summit.

---

[29] Thomas Nilsen, 'Report Encourages Norway to Reopen Olavsvern Submarine Support Base', *Independent Barents Observer*, 28 July 2016.
[30] Josh White, 'U.S. to Remove Military Forces and Aircraft from Iceland Base', *Washington Post*, 17 March 2006.
[31] *Ibid.*
[32] Trude Pettersen, 'U.S. Military Returns to Iceland', *Independent Barents Observer*, 10 February 2016.
[33] Steven Beardsley, 'Navy Aircraft Returning to Former Cold War Base in Iceland', *Stars and Stripes*, 9 February 2016.
[34] NATO, 'Strategic Concept 2010', 19 November 2010, <http://www.nato.int/cps/en/natohq/topics_82705.htm>, accessed 27 January 2017.

In August 2014, Denmark announced that it would seek to integrate the radars on its three *Iver Huitfeldt*-class air-defence frigates into the NATO ballistic missile defence (BMD) system.[35] Russia has taken an aggressive stance against Denmark joining NATO's missile defence system, stating that Danish warships would be the target of Russian nuclear missiles should it choose to do so.[36] The connection of various NATO ships with those of the US can contribute to both a more robust and mobile radar coverage.

The North Atlantic region is also home to the 12th Space Warning Squadron, which operates an Upgraded Early Warning Radar at Thule Air Force Base in Greenland. Thule contributes both as space control for NORAD and US Space Command, and performs BMD detection and tracking for the US homeland missile defence.[37] With its polar view, Thule maintains approximately 1,000 US personnel.

The US and NATO have long attempted to engage Russia on missile defence issues in the interest of transparency and mutual cooperation. In October 2013, however, Russia paused all NATO–Russia BMD-related discussions, and in April 2014, NATO suspended all cooperation with Russia in response to the Ukraine crisis.[38] NATO's missile defence capabilities will continue to strain NATO–Russia relations and could have implications for security in Northern Europe.

A New US Security Posture in the North Atlantic

Since 2014, the US has prioritised enhancing its European force deterrence posture by specifically strengthening freedom of movement and coordination between the US and its European allies. There is a heightened understanding, particularly in the US Congress, that there must be increased communication and assistance regarding the movement of troops, weapons and supplies across the Atlantic and within Europe. The US will also enhance its bilateral security partnerships in the region but these partnerships must be strengthened both ways with a recommitment to meaningfully and rapidly increased defence spending by all partners. These defence partnerships will likely engage more purposefully on the coordination of cyber capabilities and enhanced training and exercises.

---

[35] NATO, 'Ballistic Missile Defence', 25 July 2016, <http://www.nato.int/cps/en/natolive/topics_49635.htm>, accessed 27 January.

[36] Teis Jensen, 'Russia Threatens to Aim Nuclear Missiles at Denmark Ships if it Joins NATO Shield', *Reuters*, 22 March 2015; *The Local*, '"Norway Will Suffer": Russia Makes Nuclear Threat over US Marines', 31 October 2016.

[37] Jeremy Bender, 'The Most Isolated US Military Base Could Get a Lot More Important', *Business Insider*, 11 November 2014.

[38] NATO, 'Ballistic Missile Defence'.

Increasingly US naval officials and NATO commanders recognise that the Alliance must also increase its physical presence in the North Atlantic (as represented in both Norway and Iceland), increase its exercise and training tempo, enhance its intelligence, surveillance and reconnaissance (ISR) capabilities and strengthen intelligence-sharing among Allies. As part of its European Reassurance Initiative (since renamed the European deterrence initiative, or EDI), the US has requested $3.4 billion in Fiscal Year 2017, of which a portion will be used to support the deployment of 330 additional marines to Norway. This deployment is located in central Norway at the Værnes Air Station, Trondheim, which is near existing US pre-positioned equipment. In March 2016, NATO engaged in Exercise *Cold Response* in Trondheim, which involved 2,000 US marines as part of a 16,000-troop exercise which engaged fourteen countries. EDI funds will also be used to upgrade an aircraft hangar at Keflavik Naval Air Station to house short-term rotations of US P-8 Poseidon maritime patrol aircraft.[39] According to a proposed 2017 budget, the Defense Department is allocating $2.5 million for anti-submarine warfare technology. This includes providing shared, immediate and mid-range all-source analysis and situational awareness in support of US and partner operations, exercises and training in US European Command.[40] These force adjustments are in response to Russian circumnavigation flights around Iceland as well as increased submarine activity near the UK and Norway.

As NATO and the US seek to enhance their forward presence and power projection capabilities in the North Atlantic, the US must re-examine and reprioritise its existing and increasingly ad hoc security arrangements and architecture in the North Atlantic. The principal drivers of US military cooperation are its bilateral defence relationships with the United Kingdom, Norway and Denmark, along with increased engagement with Iceland and potentially with Canada. However, for the most part, these relationships have focused on crisis management, counterinsurgency and out-of-area deployments to Afghanistan and the Middle East for the past fifteen years.

The US and Norway have had an enduring bilateral defence and security cooperation whose centrepiece is based on its bilateral logistics

---

[39] The White House, 'Fact Sheet: The FY2017 European Reassurance Initiative Budget Request', press release, 2 February 2016; The White House, 'Fact Sheet: European Reassurance Initiative and Other U.S. Efforts in Support of NATO Allies and Partners', press release, 3 June 2014; Terje Solsvik and Stine Jacobsen, 'Some 330 U.S. Marines to be Temporarily Stationed in Norway in 2017', *Reuters*, 24 October 2016.
[40] US Department of Defense, Office of the Under Secretary of Defense (Comptroller), 'European Reassurance Initiative', February 2016.

planning, defence cooperation in armaments and security cooperation.[41] This cooperation includes significant defence industrial cooperation to include the Joint Strike Fighter programme (F-35) and the F-16 Multinational Fighter Program. In November 2008, the Norwegian government committed to procuring the F-35A as a replacement for its F-16 fleet. The government has funded the procurement of 22 out of a total of 52 F-35As, with subsequent authorisation occurring on a yearly basis. The first two F-35As will arrive in Norway in 2017.[42] Norway's participation in this programme will provide much needed continuity in research and development as well as technical cooperation with the US and other participants in the programme. Kongsberg Gruppen, a Norwegian technology company, is developing the Joint Strike Missile for integration into the F-35 to expand capabilities for anti-surface warfare missions.[43] Other Norwegian companies will be involved in hi-tech components for F-35 production, including advanced composites, aero structures, communications and sensors.[44]

The second component of Norway's defence modernisation is the updating of its aging P-3C Orion maritime patrol and surveillance aircraft. The P-8 Poseidon can be equipped with an array of offensive weapons, including torpedoes, cruise missiles, bombs, and mines.[45] Considering these capabilities, the P-8 will patrol the increasingly strategic GIUK Gap with the ability to monitor increased submarine activity. As its maritime surveillance capabilities have atrophied over the years, an upgrade to the P-3 shows Norway's renewed focus on security in the Northern Atlantic.

Norway's increase in defence spending in 2016 has strengthened the US–Norwegian bilateral defence relationship as has the recognition, by both countries, of the importance of forward presence and pre-positioned equipment in the region.[46] The forward-deployed US force, which will participate in exercises and training, not only shows the strength of bilateral relations, but it also enhances Norway's and NATO's ability to 'rapidly aggregate and employ forces in Northern Europe'.[47]

---

[41] US Embassy, Norway, 'Defense Cooperation', <https://norway.usembassy.gov/defense.html>, accessed 27 January 2017.

[42] Lockheed Martin, 'F-35 Partnership: Norway', <https://www.f35.com/global/participation/norway>, accessed 27 January 2017.

[43] *Ibid.*

[44] *Ibid.*

[45] Naval Air Systems Command: Aircraft and Weapons, 'P-8 Poseidon', <http://www.navair.navy.mil/index.cfm?fuseaction=home.display&key=CFD01141-CD4E-4DB8-A6B2-7E8FBFB31B86>, accessed 27 January 2017.

[46] Rebecca Kheel, '330 Marines to Deploy to Norway Amid Tensions with Russia', *The Hill*, 24 October 2016.

[47] *Ibid.*

These US forces complement existing US pre-positioned equipment stocks, as listed earlier in this chapter. This stock is also configured for munitions and aviation support equipment which allows for rapid mobilisation and global projection, thereby showing a renewed commitment to readiness in this region.[48] Finally, Norway is contributing forces to the German-led NATO battalion in Lithuania as part of NATO's Enhanced Forward Presence.

Bilateral defence relations between the US and Denmark cover a similar array of global security challenges, with a renewed focus on air and maritime cooperation in recent years. Similar to Norway, Denmark is part of the Joint Strike Fighter programme, having joined in 2002. In June 2016, Denmark confirmed plans to procure 27 F-35As.[49] Denmark has received four training and transition F-35As and beginning in 2017, six F-35As will arrive annually provided that the Danish Parliament approves funding.[50] As with Norway, Danish participation in the Joint Strike Fighter programme strengthens joint cooperation in research and development as well as technical cooperation between US and Danish industry.[51] Denmark's procurement of the F-35A will significantly enhance its and NATO's defence capabilities in the North Atlantic. Finally, Denmark is also contributing, contingent upon the Danish Parliament's approval, forces to Estonia (up to 200 soldiers) to the British-led NATO battalion as part of NATO's Enhanced Force Presence.[52]

While the US and the UK cooperate on global issues in areas ranging from the Middle East to North Africa, there has been a renewed focus on the security of Europe. In its 2015 Strategic Defence and Security Review (SDSR), the UK signalled its intent to purchase nine new maritime patrol aircraft to protect its nuclear deterrent, identify and track hostile submarine activity and enhance maritime search and rescue.[53] To date, the UK has ordered nine P-8 Poseidon aircraft, with the first to be delivered in 2019, and the remaining to be delivered within the following 24 months. The P-8s will fill an urgent capability gap created after the

---

[48] Cavas, 'Cave-Dwellers'.

[49] Lockheed Martin, 'Future Air Power: Denmark', <https://www.f35.com/global/participation/denmark> accessed 27 January 2015.

[50] *Defense Industry Daily*, 'Norway Reiterates Commitment to F-35s', 4 November 2016.

[51] Lockheed Martin, 'Future Air Power: Denmark'.

[52] *The Baltic Times*, 'NATO Leaders Decide to Deploy Battalions to Baltic States, Poland', 9 July 2016.

[53] HM Government, *Securing Britain in an Age of Uncertainty: The Strategic Defence and Security Review (SDSR)*, Cm 9161 (London: The Stationery Office, November 2015).

Nimrod MRA4 project was cancelled in 2010.[54] The SDSR also reaffirmed the UK's commitment to the F-35 programme, outlining its plan to purchase 138 F-35 aircraft over the life of the programme.[55] The acquisition of the F-35s and P-8s ensure that the US and UK will be highly interoperable, capable of landing on each other's aircraft carriers, integrating and land and sea capabilities, and improving anti-submarine warfare capabilities.[56]

The UK has observed increased Russian submarine activity as well as, in October 2016, the sole Russian aircraft carrier, *Admiral Kuznetsov*, and escort vessels traversing the English Channel en route to the Eastern Mediterranean. In response, the UK sent HMS *Duncan* and HMS *Richmond* to track the Russian carrier group.[57] The UK has also recently voted to renew its Trident nuclear weapons system as well as to build four replacement submarines.[58] HM Naval Base Clyde at Faslane, home to the core of the Royal Navy's submarine service, includes four *Vanguard*-class nuclear submarines which carry Trident II D-5 ballistic missiles as well as the latest generation of *Astute*-class attack submarines.[59]

Conclusion: Restoring NATO in the North Atlantic

An enhanced NATO presence in the North Atlantic begins with leadership combined with technologically advanced defence capabilities, a so-called Enhanced Northern Presence. This leadership begins with the four core NATO countries in the North Atlantic: the US; the UK; Norway; and Denmark. Iceland is also a key NATO Ally, despite its lack of military forces, and the government of Canada may also be engaged at a later stage of development. The four core NATO states could begin by undertaking a leadership role in providing substantial forces for Exercise *Trident Juncture*.

This 'North Atlantic Quad/Quint' have committed to increasing national defence spending and interoperability with the US. Each country is an active participant in the F-35 programme, which substantially expands power projection capabilities. The UK has purchased, and Norway is considering, P-8 Poseidon maritime patrol aircraft which will be augmented by a US P-8 rotation at Keflavik Naval Air Station. While these investments will strengthen anti-submarine warfare capabilities in the

---

[54] Nicholas de Larrinaga, 'Farnborough 2016: UK Orders P-8 Poseidon Maritime Patrol Aircraft', *IHS Jane's 360*, 11 July 2016.
[55] HM Government, *Securing Britain in an Age of Uncertainty*.
[56] *Ibid.*
[57] BBC News, 'Russian Warships Pass through English Channel', 21 October 2016.
[58] BBC News, 'MPs Vote to Renew Trident Weapons System', 19 July 2016.
[59] Stuart Nicolson, 'What Do We Know about Faslane, the Home of Trident Nuclear Weapons?', *BBC News Scotland*, 31 August 2015.

North Atlantic, improving overall capacity requires additional components and capabilities. There are also substantial anti-submarine warfare capabilities in the region through the national purchase of the Aegis weapons systems by Denmark, Norway and the UK. Finally, US and Norwegian participation in NATO's Alliance Ground Surveillance systems could have the benefit of bringing the Global Hawk unmanned aerial vehicles capabilities to the North Atlantic. There are also growing integrated air- and missile-defence capabilities at sea as the UK and Norway both participate in the Maritime Theatre Missile Defence Forum.

In recognition of significant military capabilities in the North Atlantic, there is an opportunity to enhance regional command and control functionality. While it is neither feasible nor efficient to reinstate a US Atlantic Command, consideration should be given to considerably strengthening the 300-person NATO Allied Maritime Command (MARCOM) in Northwood, in the UK, and double-hatting it as a joint NATO and UK North Atlantic Command. Bilaterally, the UK could initiate activities and exercises by the North Atlantic Quad. This could augment existing NATO activities and facilitate participation by other Allies such as France. A more robust MARCOM and Quad arrangement could increase NATO Allies' maritime exercise and training in 2017 that would build towards *Trident Juncture*, which should include up to 25,000 NATO forces. If possible, this exercise should have a maritime component that would include exercising anti-submarine warfare capabilities, minesweeping and the detection of underwater drones.

Combined, a more focused MARCOM supplemented by the North Atlantic Quad leadership would represent measurable steps to restore NATO's presence and capabilities in the North Atlantic.

IV. NORWAY AND THE NORTH ATLANTIC: DEFENCE OF THE NORTHERN FLANK

SVEIN EFJESTAD

After 9/11, NATO invoked Article V and declared its readiness to come to the assistance of the US. This was a type of Article V engagement very different from any foreseen in 1949 or during the Cold War. NATO engaged in the surveillance and protection of the US homeland and in military operations in Afghanistan to prevent that country from remaining a safe haven for international terrorism. The Alliance had already assumed new roles and tasks in the 1990s, including crisis management and cooperative security, as described in the Strategic Concept of 1999.[1] Collective defence, however, is the only task directly based on the 1949 North Atlantic Treaty and thus binding on all member states.

NATO adapted to the new geopolitical reality and the concomitant emphasis on counterinsurgency operations and peace enforcement, but at the expense of its ability to exercise collective defence. With Russia resurgent, increasingly challenging and testing Western interests and values, NATO needs to adapt yet again in order to regain an effective collective defence posture. Credible defence and stability in the North Atlantic will create a sound basis for NATO's defence and deterrence elsewhere in Europe, and effective transatlantic collective defence in Europe depends on protecting the sea lines of communications across the Atlantic.

---

[1] NATO, 'The Alliance's Strategic Concept', NAC-S (99) 65, press release, 24 April 1999, <http://www.nato.int/cps/en/natolive/official_texts_27433.htm>, accessed 25 January 2017.

This chapter argues that NATO needs a comprehensive approach for its most important maritime domain: it must strengthen its command and control structure, invest in high-end capabilities, update its contingency plans and change its exercise policy in accordance with the new security reality.

From Cold War Confrontation to Out-of-Area Operations

Towards the end of the Cold War, in the late 1980s and the beginning of the 1990s, NATO had an impressive, integrated military defence structure and well-rehearsed decision-making procedures that could take advantage of the superior forces supplied by the member states. The most important elements of this collective defence structure were common defence planning with an emphasis on capabilities and operations, large-scale exercises, fit-for-purpose command arrangements, and a relevant and functioning decision-making apparatus. NATO had extensive command arrangements in the 1980s.[2] The entire structure participated in operational planning, exercises and preparations for crisis management and armed conflict vis-à-vis the Soviet Union and the Warsaw Pact countries. The Alliance had a capacity to provide good situational awareness and advice and it offered recommendations through the chain of command up to the NATO Council.

The interface between NATO's European Command and the Atlantic Command in Norfolk, Virginia, was complex; the border between their areas of responsibility ran along the Norwegian coast. Operations across the command boundaries had to be carefully planned and communicated. At times, this represented a great challenge for the two joint commands in Norway which were double-hatted national commands in peacetime, but ready to be transferred to NATO in times of crisis or war. The commands in Norway belonged to Allied Command Europe, but cooperated closely with Allied Command Atlantic.

In the last phases of the Cold War, NATO had extensive plans for the employment of forces in all parts of Europe.[3] These plans covered the deployment of reinforcements to exposed areas and arrangements for host-nation support as well as preparations for the rapid transfer of command to a designated NATO commander.

---

[2] Jacob Børresen et al., *Norsk forsvarshistorie, bd. 5, Allianseforsvar i endring, 1970–2000* [*The History of Norwegian Defence, Vol. 5, The Changing Character of Alliance Defence 1970–2000*] (Bergen: Eide forlag, 2004), pp. 85–105.
[3] *Ibid.*; Jacob Børresen, 'Alliance Naval Strategies and Norway in the Final Years of the Cold War', *Naval War College Review* (Vol. 64, No. 2, 2011), pp. 97–115.

In Northern Europe, these arrangements changed when NATO closed the Northern Command in Oslo responsible for Schleswig-Holstein, the Danish straits and Norway in 1994. Norway and the UK then formed the new Northwest Command but this headquarters closed in 2000. The joint operational command in Brunssum took over the responsibility for that part of Europe. Norway kept a joint sub-regional command in Stavanger that was a mixed national and NATO command until it closed in 2004. Since 2009, Norway has had one single joint operational command located in northern Norway, but it has no direct relationship with NATO.

The comprehensive command reform in 2003 was heavily influenced by the focus on Afghanistan and Iraq. The closure of the Atlantic Command in Norfolk in 2003 was a game-changer. Allied Command Transformation (ACT) assumed responsibility for training and exercises as well as force planning in the NATO structure. Again, the priority was to assist in the development of deployable forces for counterinsurgency operations. The NATO summit in 2010 decided to reduce the Command Structure again and consequently there was less capacity to prepare for collective defence.

After the Soviet Union had collapsed, NATO encouraged the dissolution of mobilisation forces and the elimination of conscription. Planning now reflected that the forces would be used in counterinsurgency operations, peace enforcement and peacekeeping. Moreover, NATO's planning system did not encourage investment in air defences and maritime forces – including submarines – or in heavy army forces with significant firepower. Instead, the demands of ongoing operations dominated planning. Readiness and maintenance were adjusted to meet the requirements of the next rotation of forces in the theatre of operations, and logistics focused on the current mission. Training and exercises were designed to prepare the forces for the challenges they could expect to counter in action.

This orientation has continued for at least 20 years, and the results are still reflected to varying degrees in the status of forces in most Western countries. It has led to a lack of sustainability, readiness and survivability and to a shortage of forces designed and trained for high-intensity warfare. Furthermore, the link with national civil emergency organisations became weak, and generally the plans for civilian support to the armed forces in crises and war were not kept up-to-date and exercised.

Many member states also shelved their contingency plans. As time went by, these plans became largely irrelevant because military units and commands were disbanded and support institutions and arrangements disappeared. The exercise pattern prevalent in the Cold War also disappeared. Exercising the defence of NATO territory became controversial. Moreover, traditional defence exercises were expensive and no longer received support from

NATO's common funding. The operational tempo required for deployed operations did not allow for the extensive use of forces for other purposes. After two decades of focusing on deployed counterinsurgency operations, the forces and the new generation of senior military officers had developed an operational focus very different from the traditional NATO approach which sought to maximise deterrence.

Norwegian Initiative in 2008: Raising NATO's Profile

For a small country in a strategically important location, collective defence takes on a meaning quite different from that understood by a great power. The Norwegian national forces are not an independent defence force able to fight a high-intensity war on their own; they must be integrated into a larger framework. They can, however, handle smaller crises and incidents on their own and they can prevent a superior opponent from taking control without resorting to massive use of military force. An effective collective defence concept requires a designated, well-qualified command arrangement and plans that integrate specific forces into a framework for defence. If NATO cannot provide such a framework, smaller countries might be tempted to develop their forces in a more traditional direction or to enter into bilateral arrangements outside the established organisation. Thus, a weak NATO could pave the way for the further renationalisation of defence planning and create a self-reinforcing spiral that undermines NATO's efficiency in the longer term.

At an informal defence ministerial in London in June 2008, the then Norwegian minister of defence, Anne-Grete Strøm-Erichsen, presented a paper on raising NATO's profile in the member states.[4] The main agenda was to better balance 'out-of-area' operations with 'in-area' activities, acknowledging that there was an urgent need to revitalise the key concept of deterrence and collective defence. She argued that NATO needed improved situational awareness and better linkage to the national military commands, and that NATO's military organisation should be more involved in training and exercises. The paper highlighted a sense that NATO was drifting in a direction where its relevance to the defence of the member states was becoming questionable. At that time, political and military developments in Russia clearly did not signal a move towards greater cooperation with the West. The war in Georgia started two months later.

---

[4] Paal Sigurd Hilde and Helene Widerberg, 'Norway and NATO: The Art of Balancing', in Robin Allers et al. (eds), *Common or Divided Security: German and Norwegian Perspectives on Euro-Atlantic Security* (Frankfurt am Main: Peter Lang, 2014), p. 199.

Following the Norwegian paper, NATO decided to strengthen its links to the military establishments in member states, but progress was extremely slow and NATO's ability to implement changes in its military activities proved disappointing. The increased number of member states, each with a different threat perception, bureaucratic traditions and specialties, had a negative impact. This experience demonstrated the need for reform and for greater transparency and oversight to improve NATO's effectiveness. However, some changes proved positive. Supreme Headquarters Allied Powers Europe, for example, built a new situation centre that has proven useful in providing situational awareness and distributing relevant information.

Dialogue and Cooperation with Russia

After the end of the Cold War, it became apparent that the Russian armed forces and the infrastructure supporting Russia's military establishment had severe shortcomings. Russia's defence industry was kept running due to orders from foreign countries. In some cases, it developed capabilities that Moscow could not afford to buy. During the 1990s, Russia demonstrated unprecedented transparency and cooperation, and NATO executed the majority of its out-of-area operations in a relatively cooperative environment, although there were exceptions: Operation *Allied Force* – NATO's campaign of air strikes in Kosovo – in 1999 did not receive an explicit mandate from the UN Security Council. This NATO operation caused much bitterness and resentment in Russia, clearly demonstrated during the visit of the then Russian Defence Minister Igor Sergeyev to Stavanger in 1999. Tensions between Russia and NATO continued and became very visible during the operations in Kosovo. Nonetheless, Russia saw the operations in Afghanistan from 2001 as a fight against terrorism and supported them more or less directly. Russia had to confront its own domestic terrorist challenge and suffered several brutal attacks against its civilian population, waging two wars in Chechnya to re-establish Russian control.

At the bilateral level, Norwegian–Russian defence cooperation had been established in 1992 when Johan Jørgen Holst, the Norwegian minister of defence at the time, visited Russia. The ministers agreed to conduct joint naval exercises and other NATO members operating naval forces in the North were also invited to participate. The first such exercise, *Pomor*, was held along the Norwegian coast in 1994. In 1995, Russia and Norway established a dialogue on defence and security, which resulted in numerous high-level visits, seminars, training and exercises, and a routine dialogue on security and defence. The two ministries of defence also had a yearly programme of cooperation and contact.

The ministers of defence of Russia, the US and Norway entered into a cooperative agreement to prevent nuclear pollution in the Arctic in September 1996. This programme (Arctic Military Environmental Cooperation) continued with good results for many years. This historical moment of 'deep peace' was not to last. Russia's highest priority, particularly in the High North, was to keep its nuclear forces in a state of readiness so that it would still be recognised as a nuclear superpower. As soon as Russia started to recover, in part because of high oil prices, it started to reinvest in conventional military forces. Gradually, Russia became more critical of the Western powers and initiated a more authoritarian and nationalistic domestic policy. This also had implications for Russia's relations with the West, as did Moscow's restrictions on the activities of Western organisations and institutions in Russia.

The New Russian Posture

The 2008 war in Georgia represented a major setback for Moscow's relations with NATO, and demonstrated the many deficiencies in the Russian armed forces. As a result, the Russian government launched a massive effort to modernise the Russian armed forces, placing emphasis on new equipment, readiness, and command and control. From 2008, Russia started to take a more forward-leaning stand both politically and militarily.

President Vladimir Putin was adamant about re-establishing Russia as a great power. With the State Armaments Program 2020, Russia set very ambitious targets for its armed forces: in particular, a goal of replacing 70 per cent of its weapons systems with new ones.[5] A sharp increase in defence spending resulted in new weapons systems, higher readiness and a new extensive training and exercise pattern. Although the decline in oil prices from 2014 and other factors have since forced it to moderate its ambitions, Russia still gives very high priority to its armed forces and their modernisation.

Russia inaugurated a new National Command Centre on 1 December 2014, and has used it during the most recent large-scale exercises, such as Exercise *Kavkaz 2016*. This centre, which has authority over all sectors of government, would serve as a key instrument in shifting Russian society from a peacetime to a wartime posture. Given the extreme concentration

---

[5] Tor Bukkvoll, 'The Russian Defence Industry – Status, Reforms and Prospects', Norwegian Defence Research Establishment, FFI report 2013/00616, 3 June 2013; Una Hakvåg et al., '*Skremmende tall: Realismen i det russiske våpenprogrammet GPV-2020*' ['Scary Numbers: The Realism in the Russian Weapon Programme GVP-2020'], Norwegian Defence Research Establishment, FFI report 2012/00356, 4 September 2012.

of power in Russia, such a set-up would enable it to act rapidly and comprehensively.

Russia conducts snap exercises to check the readiness of its forces, which has improved greatly over the past few years. This is true also for the forces in northwest Russia and the Northern Fleet. The exercises demonstrate that force mobility has also improved, giving Russia the option to deploy its forces at short notice.

Moscow has stated that Russia needs a military capability in the Arctic to protect its interests there, and has started a substantial build-up of military forces in the region, particularly in the Kola Peninsula and surrounding areas. The Russian press reports that Russia will station new Su-34 multirole aircraft and air-refuelling planes permanently at Franz Josef Land.[6] This and numerous other projects will improve the Russian military coverage of the European Arctic, and could also be an important stepping stone for establishing military activities further west into the Atlantic and the Norwegian Sea.

On 1 December 2014, Russia also established the Arctic Command.[7] Putin stated that it would be responsible for defending Russia's interests in the region in collaboration with the Northern Fleet, thus creating a joint command stretching from the Arctic to the Eastern Atlantic.

Russia keeps strategic submarines (*Delta IV* and *Dolgorukiy* class) on patrol in the Northern waters whose missiles provide strategic deterrence. Russia's long-range strategic aircraft (Tu-95 Bear and Tu-160 Blackjack) have modern cruise missiles and provide the air component of the nuclear triad. Flights across the North Pole and along the coast of Western Europe demonstrate readiness and capability. Modern attack submarines (the *Severodvinsk* and *Akula* class) deploy into the Atlantic to protect the strategic assets in the North. They also have substantial capabilities for long-range precision strikes against land and sea targets. New long-range coastal defence missiles are being deployed in the North. Russia plans to strengthen its surface fleet, but that fleet is still limited in numbers and consists largely of older ships.

It is unlikely that a military conflict between Russia and NATO would start in the High North or in the Northern Atlantic. There will be overlapping claims on the continental shelf in some areas, but there are no reasons to believe that such issues will lead to serious conflict.[8] But if a military

---

[6] Atle Staalesen, '700 Men Building New Airfield in Franz Josef Land', *Barents Observer*, 26 October 2016.

[7] Trude Pettersen, 'Russian Arctic Command from December 1st', *Barents Observer*, 25 November 2014.

[8] For more on the Arctic Council, see <http://www.arctic-council.org/index.php/en/>, accessed 25 January 2017.

confrontation between Russia and the West starts in other areas, there is every reason to believe that it will have direct and immediate consequences for the situation on the northern flank, where Moscow has concentrated much of its strategic military capabilities. Still, cooperation between the West and Russia is probably better in this part of the world than in most other areas. Although Russian actions in Ukraine caused a complete halt in military cooperation, Norway and Russia have worked together in the High North on issues of mutual concern such as fisheries, coast guard activities, confidence-building and border control. The Arctic Council remains a useful instrument for promoting cooperation in environmental protection, search and rescue, and polar research.

A New Direction in Allied Defence Planning

Should Article V of the NATO Treaty be triggered, military operations could take place anywhere in Alliance territory; long-range, precision-guided weapons and mobility make it most unlikely that such operations will be restricted to a limited geographic area. This situation, and the need to use Allied forces with great flexibility, demands a comprehensive joint combined operational plan for the defence of Europe. Although it is impossible to predict how a large military conflict in Europe might start and how it might develop, such a conflict would probably encompass cyber attacks, hybrid elements and degradation of infrastructure in addition to the use of conventional military forces. Resilience will depend to a large degree on solidarity among all members of NATO, and on the willingness and ability of the US to reinforce Europe if needed.

NATO members agree that collective defence remains the Alliance's most important task, although crisis management and cooperative security have become, and will remain, important and demanding. Re-establishment of credible collective defence must be the priority, but the Alliance must also sustain crisis management operations far away from Europe. To achieve this, NATO needs more resources, as heads of states and governments decided at the Wales Summit in 2014. The UK, Germany, Norway and others have already increased their defence budgets and committed themselves to greater investment in capabilities for collective defence. The Readiness Action Plan, the European Reassurance Initiative and the improvement to the NATO Reaction Force represent concrete examples.[9] These and other initiatives have put NATO in a much better position to deter, defend and reassure Allies than before the NATO summit in Wales.

---

[9] See, for example, Terri Moon Cronk, 'European Reassurance Initiative Shifts to Deterrence', *DoD News, Defense Media Activity*, 14 July 2016.

At present, most of NATO's measures are reactive in character and intended directly to reassure those countries that feel most exposed to the renewed Russian military assertiveness. This is true for the Enhanced Forward Presence initiative and the reinforced air policing in the Baltics. While such actions are both positive and necessary, NATO must adopt a more holistic approach to re-establish its primacy as an anchor of stability and security for the Euro-Atlantic area. Such an approach could build greater unity and strengthen defence and deterrence with only moderate increases in defence spending, and could better integrate the relatively small contributions that most European countries can make to NATO's overall posture.

A more comprehensive approach to collective defence should start with a realistic planning effort for joint and combined operations in NATO's different geographic areas. Credible deterrence includes preparedness to escalate from peacetime to a crisis posture and if necessary to a full military defence effort. Participants in this planning effort must discuss the basic assumptions and challenges that NATO could face. The analysis should address topics such as the need for in-place forces, readiness, reinforcement requirements, logistics support, host-nation support and force protection, training and exercises and arrangements for the rapid establishment of effective unity of command.

NATO developed a comprehensive approach to peace enforcement in Afghanistan; it must now redevelop a similar 'whole-of-government' approach to collective defence. NATO's resources remain superior to those of any conceivable opponent and the Alliance should incorporate this advantage into well-prepared, cohesive operational planning. NATO has an extensive system capable of preparing and coordinating such an effort, and its posture becomes much stronger when contingency planning for joint operations better incorporates follow-on forces.

The defence of the North Atlantic and surrounding territories is vital for transatlantic cooperation in crises and war, and therefore for the credibility of collective defence in all geographical areas of NATO. The Alliance must be able to use bases in Norway, Iceland and the UK in order to exercise sea-control and dominance in the Eastern Atlantic. The area north of the Greenland-Iceland-UK Gap and the Norwegian Sea, including the western part of the Arctic where military activities have again increased, presents a major concern. Preparing for the summit in Warsaw, the Norwegian Ministry of Defence, in cooperation with the UK, France and Iceland, launched new proposals aimed at strengthening the Allied posture and activities in the North Atlantic.

NATO should bolster existing forces by building adequate readiness, survivability and sustainability. This will require a sustained and expensive effort. NATO must rebuild wartime stocks of ammunition, fuel, spare parts and other supplies; invest more in survivable military infrastructure; and establish a well-trained personnel reserve. The Alliance must also revitalise its total defence planning, which means clarifying what goods and services the military would need from the civilian sector. This will prove demanding, because all planned functions are based on the assumption that data networks and modern communications will remain available, and storage of goods has been reduced to a minimum to save money. Thus, NATO must create a new system for emergency support to civilian populations – a very demanding task given the vulnerability of modern societies.

After the Cold War ended, NATO members almost neglected national planning for crises and war, and for the protection of the civilian population in exposed areas. Instead, they focused on military planning for support to the police in different emergencies, not least counter-terror operations. In Norway, this changed around 2010 when the government decided to launch a new effort, primarily because it realised that preparations for the defence of the country's territory and population were inadequate. The new effort included updating contingency plans, readiness requirements and preparations for logistics support. Russian exercises conducted with little or no warning confirmed the requirement for improved readiness. The Norwegian government also underlined the importance of restoring civil preparedness and the total defence concept. This is important not only for the support of Norwegian forces and Allied reinforcements, but also for protecting the civilian society in emergencies.

Gradually, authorities in many countries have come to recognise a need to revitalise such arrangements, and have already responded to the new security situation by implementing various internal measures. The protection of civilians requires overall coordination of military operations and understanding of their implications for civilian society – an understanding that NATO currently lacks. Therefore, such preparations must be a national responsibility, with member states setting up integrated civilian–military crisis management organisations, and NATO must contribute general guidelines to help civilian societies prepare for crises or war.

The Broader Nordic Dimension

The Nordic countries' security orientation has changed considerably over the past decade. When Sweden and Finland joined the EU in 1995, they were included in the European security community but were still not part of NATO's collective defence: they cooperated closely with NATO but remained non-aligned. At the Wales Summit in 2014, they became

Enhanced Opportunities Partners.[10] This new status has given both countries better access to NATO, as demonstrated at the Warsaw Summit.[11]

Close cooperation between NATO and Sweden and Finland strengthens the Alliance's deterrence in Northern Europe. While these countries have no defence guarantee from NATO, their capabilities and participation play an important role. This Nordic cooperation provides links to the Baltic region, and thus to the security of the North Atlantic. Sweden and Finland have been involved in NATO's deliberations on the security of the Baltic Sea area and both countries have entered into agreements with the US and the UK, paving the way for comprehensive bilateral defence cooperation.[12] The Nordic Defence Cooperation (NORDEFCO), which encompasses joint or coordinated contributions to international operations, has served as a platform for improved military cooperation between Sweden and Finland and members of NATO.[13]

The Nordic countries have agreed that their air forces can operate freely across national borders in training and exercises. This agreement includes the airspace of all five Nordic countries. In Northern Scandinavia, the air forces of Finland, Sweden and Norway conduct cross-border training exercises on a weekly basis.[14] The Nordic countries have arranged relatively large air force exercises (*Northern Challenge*), with Allied countries also taking part. Finland and Sweden have participated with large forces in the large biennial *Cold Response* exercises in Norway involving all major NATO countries; these exercises have occasionally included Swedish territory. Finland and Sweden have achieved a very high degree of interoperability with NATO, enabling joint operations at short notice, and both countries have made substantial contributions to

---

[10] NATO, 'NATO Secretary General Welcomes Deepening Cooperation and Dialogue with Finland and Sweden', 1 December 2015.

[11] NATO, 'Wales Summit Declaration', (2014) 120, press release, <http://www.nato.int/cps/en/natohq/official_texts_112964.htm>, accessed 25 January 2017; NATO, 'Warsaw Summit Communiqué', (2016) 100, press release, 9 July 2016, <http://www.nato.int/cps/en/natohq/news_125372.htm>, accessed 25 January 2017.

[12] For Finland, see NATO, 'Relations with Finland', 28 November 2016, <http://www.nato.int/cps/en/natohq/topics_49594.htm>, accessed 26 January 2017; and for Sweden, see NATO, 'Relations with Sweden', 28 November 2016, <http://www.nato.int/cps/en/natolive/topics_52535.htm>, accessed 26 January 2017.

[13] The main aim and purpose of NORDEFCO is to strengthen the participating states' national defence, explore common synergies and facilitate efficient common solutions. It is a cooperation structure, not a command structure. Cooperation activities initiated from the top or bottom are facilitated and agreed within the structure, but the actual realisation and participation in activities remain national decisions. See NORDEFCO, <http://www.nordefco.org/>, accessed 26 January 2017.

[14] Swedish Ministry of Defence, 'NORDEFCO Annual Report 2015', January 2016, p. 9.

NATO-led operations in the Balkans and in Afghanistan. Sweden and Finland contribute to NATO's Reaction Force and may also be involved with other Western force pools in the future.[15]

Priorities for NATO

The present, demanding global security environment features many competing requirements, forcing NATO and its member states to find cost-effective and flexible solutions. These arrangements must promote cohesion among all of NATO's members. NATO can accomplish this only if it tailors its footprint to the different circumstances prevailing in different regions. A general contingency plan would create a better point of departure than focusing on the number of operations that NATO should be able to conduct at any given time. Moreover, most of NATO's forces are deployable and can be used in contingencies outside Alliance territory. It is not realistic to earmark forces for specific areas or specific contingencies, because NATO simply does not have enough forces. Even so, NATO might find it useful to identify capabilities and forces for possible deployment to different areas. This would facilitate preparation for the force provider and for the receiving (host) nation, and would aid in the preparation of command arrangements.

Such a strategy would also enable NATO to establish an improved training and exercise programme. At present, one or more countries arrange large exercises without coordination inside NATO. *Cold Response*, *Arctic Challenge* and the Baltic Sea *Baltops* exercises are examples. This reduces NATO's visibility, deprives it of a basis for learning lessons and incorporating them into its operational activities and planning and reduces the Alliance's options for using exercises to qualify its staff to manage crises and conduct operations.

Double-hatting would be an effective way to overcome these shortcomings without increasing personnel. Double-hatted headquarters could carry out various tasks, including contingency planning, exercises, crisis management and the conduct of operations. Double-hatted headquarters were used with good results during the Cold War, for example in Norway, Denmark and the UK. They could also help identify shortfalls and develop proposals for new investments. Many large joint and combined exercises revealed shortcomings and inadequate capabilities, and these findings were later used in NATO's capability planning. National joint operations headquarters could organise large exercises on behalf of NATO and report to Allied Command Operations.

---

[15] NATO, 'Sweden to Join NATO Response Force and Exercise Steadfast Jazz', 14 October 2013; NATO, 'Relations with Finland', 28 November 2016.

These headquarters can be transferred to NATO command in actual crises, thereby greatly extending SACEUR's capacity, competence and situational awareness. NATO could also do more to prepare a pool of well-qualified officers who could reinforce headquarters that may be particularly exposed in certain emergencies, thereby improving NATO's organisation at low cost. The 2016 summit in Warsaw decided that NATO would conduct a functional assessment of the current command arrangements. This was the result of a Norwegian initiative.

Not all member states need to participate in all activities at all times; the inclusion of many new member states with limited resources has made this unrealistic and ineffective. Therefore, NATO must improve its ability to develop policy and guidance for all its members, and give greater responsibility for implementation to those states that have active military engagement in the different regions. This would strengthen NATO's situational awareness and overall effectiveness, and ultimately improve its deterrence and defence posture. If NATO applies these principles in the North Atlantic, it would have one joint headquarters with primary responsibility for this area. The headquarters would carry out contingency planning, organise exercises and prepare itself to manage crises.

One possibility would be to create a new strategic command similar to the old Supreme Allied Command Atlantic. Two strategic operational commands would require more coordination. A better solution might be to build on the maritime command in Northwood and make the North Atlantic its primary area of responsibility. In this case, the Northwood command should be a joint operational command but focused on the maritime domain, primarily manned from regional states or operating in the North Atlantic. Several national headquarters have the appropriate competence and capability to assist this command. With maritime headquarters in the US taking responsibility for the Western Atlantic, the joint operational headquarters in Norway and perhaps national headquarters in Denmark and the UK could form a web of highly capable headquarters in the North Atlantic.

The new headquarters could act as a hub for military coordination and integration. From a political perspective, such arrangements would give member states a sense of ownership and confidence in NATO's collective defence. The regional hub would also strengthen the links between NATO and its member states.

It is difficult to determine what force level would suffice for effective deterrence and defence in the North Atlantic area. All countries directly engaged in the defence of the region are now increasing their investment in relevant capacities. The investment should focus on maritime and high-intensity warfare capabilities, with particular emphasis on improvements in air defences and anti-submarine warfare.

Exercise policy should aim to ensure NATO's collective capacity to establish sea-control to ensure that Allied reinforcements and supplies can flow across the Atlantic to Europe. Exercises should also incorporate elements of civil defence. NATO need not conduct these activities close to the Russian bases at Kola. The Alliance should offer Russia transparency and predictability. Although Russian military activities stretch further north and west, and its snap exercises give no warning and no transparency, it would not be in NATO's interest to participate in a cycle of destabilisation. Instead, NATO should continue to try to convince Russia to show restraint and increase transparency.

Implications for Norway's Defence

For a small state such as Norway, a credible and effective NATO forms the basis for safeguarding its national security and constitutes the foundation for contacts and cooperation with Russia. Norway recognises Russia's legitimate security interests in the High North, but a firm and realistic security and defence posture should not have a negative effect on relations with Moscow. At the same time, Norway realises it must not contribute to an atmosphere of military confrontation or conduct any activity in the border areas with Russia that could reasonably be interpreted as provocative. Therefore, Norway does not allow large Allied exercises in the county of Finnmark, in the northeastern part of the country, and strictly controls Allied activities on Norwegian territory, particularly in the north. These restrictions in peacetime include limits on Allied air operations from Norwegian bases close to the Russian border. The policy and practice of restraint is particularly important in the current security situation.

The Norwegian government presented the new Long Term Defence Plan (White Book) to the Parliament in June 2016.[16] The proposal reflected the recognition that Norway will always depend on NATO reinforcement in crises and war, and that Norwegian military capabilities should be tailored to fit into a collective defence effort. Approved by Parliament in November 2016, it highlights the need for increased readiness, sustainability and logistic support for the present force structure. The plan implies a stronger presence in the most exposed areas

---

[16] Norwegian Government, '*Kampkraft og bærekraft – Langtidsplan for forsvarssektoren*', Prop. 151 S (2015–2016), June 2016; Norwegian Ministry of Defence, *Capable and Sustainable: Long Term Defence Plan* (Oslo: Norwegian Ministry of Defence, 2016), <https://www.regjeringen.no/globalassets/departementene/fd/dokumenter/rapporter-og-regelverk/capable-and-sustainable-ltp-english-brochure—print.pdf>, accessed 18 November 2016.

in the north and investment in high-end capabilities. This will not, however, preclude Norway's future reliance on NATO support to maintain a credible deterrence and defence posture in the north.

Norway plans to buy 52 F-35 combat aircraft, which will operate from bases along the coast. They will provide substantial airpower and their modern sensors will greatly improve situational awareness. Norway also plans to purchase five new P-8 maritime patrol aircraft to replace its ageing P-3 Orions. The five new frigates will receive more personnel in order to improve their operational capacity. The government will acquire new modern submarines to replace the *Ula* class, and Norway will maintain a modern fleet of ocean-going coast guard vessels as an integrated element of its navy.[17] The Norwegian Intelligence Service, which cooperates closely with Allied services and helps Oslo to maintain good situational awareness in the High North and the Norwegian Sea, will receive more resources. In addition, Norway has recently strengthened its special forces, and has proposed a manning level of 38,000 for the Home Guard, with about 3,000 of them trained for rapid deployment to exposed areas. Norway must revitalise its total defence concept and improve the readiness, survivability and sustainability of its forces as well as its ability to provide host-nation support.

The Long Term Defence Plan also concluded that the Chief of Defence has to conduct a study to determine the composition, tasks and organisation of the land forces. Norway intends to increase the number of troops stationed at the Russian border to two companies. The Norwegian armed forces will improve readiness, sustainability and logistic support in the near future as high-priority measures. The forces are organised and structured to act as a component of a larger NATO military framework, but the Alliance has not maintained and developed this framework in accordance with the new security situation. There are efforts to establish new plans that will facilitate a joint Allied defence and deterrence posture in the north. In the longer term, Norway will have advanced and modern forces in all services, optimised for deterrence and high-intensity warfare in the northern region.

Conclusion

NATO's defence and deterrence posture in the Atlantic Ocean is essential for the credibility of transatlantic security cooperation. Reform of the command arrangements, contingency planning, maritime strategy, training and exercise policy and greater investment in relevant capabilities for

---

[17] *Ibid.*, p. 63.

high-intensity warfare are required. The allies in the area should improve their cooperation and coordination. An Atlantic command could form a hub for discussion of political and military issues relating to defence in this area. The objective is to maintain deterrence, stability and security in the area and this will have a positive impact on the security situation in the rest of Europe. Security in the Atlantic is what ties the US and Canada directly to European security and this has to be a seamless connection.

The defence and security posture in Norway is essential for NATO's ability to operate in the northwestern part of the Atlantic and to establish a sufficient degree of sea-control to allow reinforcements and supplies to be shipped from North America to Europe in crises or war. Norway will do its part, but better overall coordination and command is required in order to get the full benefits of collective defence.

All countries in this area profit from cooperation in the Arctic. The Western approach to security in the Atlantic should not cause more tension or misunderstanding. However, NATO's policies must be predictable and firm. The objective is security and defence, not confrontation.

V. THE UK AND THE NORTH ATLANTIC: A BRITISH MILITARY PERSPECTIVE

PETER HUDSON AND PETER ROBERTS

Great Britain's naval forces have considered the North Atlantic to be a critical sphere of influence for centuries. Yet Britain never recognised this by dedicating a fleet to it – unlike the Baltic, Mediterranean, Indian or Pacific Ocean fleets.[1] While this may appear strange, it represents the centrality of the North Atlantic to UK thinking: control, freedom and access were assumed in the formulation of national, or grand, strategy. It is perhaps a sign of a continuation of such a subconscious presumption of authority that recent UK national security strategies and defence reviews have made no mention of the North Atlantic as a region of strategic interest; the implication being that free trade and travel through this ocean would be uncontested. This, in fact, has never historically been the case. Yet the orthodox view, in British minds at least, is that the Royal Navy and its allies always retained sufficient capability advantage to meet challenges to hegemony.

Few mariners would describe the North Atlantic as a homogeneous environment. Operating there effectively – above, on and below the waves – has always been difficult, demanding specialist skills, robust equipment and intellectual resilience to succeed. The Royal Navy has an excellent track record and can be proud of its achievements. These include its actions in the Battle of the Atlantic during the Second World War, and during the Cold War, when it championed innovative tactics and bold military thinking to safeguard essential resupply routes and contain a strong Soviet Navy. No more. The end of the Cold War heralded a decline

---

[1] See N A M Rodger, *The Command of the Ocean: A Naval History of Britain 1649–1815* (London: Penguin, 2004).

in blue-water capability which continues to this day. At a time when the Atlantic is once again central to the integrity of NATO and is a strategically important operational space for a resurgent Russia, the Royal Navy is poorly placed to fulfil those subconscious assumptions of dominance which shaped the DNA of the service for centuries.

This chapter will seek to explain a British maritime view of the North Atlantic by reflecting on the recent military priorities, using the lenses of capabilities, platforms, training and experience, command and control and doctrine to understand the changes that have taken place. It does so in a historical context, establishing the normalised view of Cold War activities. It then looks at the subsequent 20 years between 1991 and 2011 as the Royal Navy transformed itself from a force designed for sea-control in the North Atlantic to one that contributes to a wider expeditionary force.[2] The remaining capabilities are focused on delivering effect on land, often well away from historic areas of influence, and invariably against weak, non-complex maritime adversaries. Finally, the contemporary security environment will be examined, seeking to understand how the Royal Navy and NATO could confront the renewed tensions permeating the Atlantic and proffering thoughts and recommendations for the future.

In researching this chapter, the authors interviewed serving and retired naval officers, policymakers and scholars from around the world, many of whom did not want their evidence attributed. The authors' own opinions were shaped by their own experiences as serving Royal Naval officers in the North Atlantic over the periods under consideration: these could be considered biased and partial. As such, they followed methodologies of self-reflexivity in their own views,[3] and elite interviewing in the conduct of the external inputs.[4]

A British Military View on the North Atlantic, 1945–91

At the end of the Second World War, the criticality of the North Atlantic to UK and European security needed no explanation. From the Battle of the

---

[2] British naval doctrine defined sea-control as 'The condition that exists when one has freedom of action to use an area of sea for one's own purposes for a period of time and, if necessary, deny its use to an opponent. Sea control includes the airspace above the surface and the water volume and seabed below'. See Royal Navy, *BR 1806 British Maritime Doctrine: Second Edition* (London: The Stationery Office, 1999), p. 232.

[3] See Helena Carreiras et al. (eds), *Researching the Military* (London: Routledge, 2016).

[4] Philip Davies, 'Spies as Informants: Triangulation and the Interpretation of Elite Interview Data in the Study of the Intelligence and Security Services', *Politics* (Vol. 21, No. 1, 2001), pp. 73–80.

Atlantic to the eventual defeat of Germany, the North Atlantic was the crucial sea route to the US that kept Britain in the battle and ultimately ensured victory. Indeed, such was the seminal importance of the ocean that it was chosen as the first two words of the strategic military alliance – spawned from the wreckage of the Second World War – that has ensured European peace for over 60 years: the North Atlantic Treaty Organization. NATO's plans during the Cold War to deter the Warsaw Pact relied on unfettered access to the North Atlantic to enable the rapid reinforcement of ground forces in Germany (known as *Reforger*) with US troops. NATO Allies such as Norway depended upon reinforcement from the Atlantic for their national survival in time of conflict; much effort was made to finesse resupply operations in the country's harsh north.

The Soviet Union recognised the importance of the Atlantic as well as the High North; it was seen as a critical environment through which it could challenge NATO military efforts. In addition, it was the area of operations for its own submarine-based nuclear deterrent force (giving rise to the concept of 'the bastion' as a system of layered defence around the Arctic). It also provided access for Russian goods, trade and influence among like-minded partners such as Cuba and countries around the Mediterranean, including Libya, Egypt and Syria. As an ocean of strategic importance to both sides, there was a state of constant competition and tension in the Atlantic. As their geographic lines of influence and political-military ambition became clearer, so did their maritime strategies. In typically asymmetric fashion, NATO's focus on deployment and resupply naturally generated a Soviet desire to interdict it through extensive submarine activity and coordinated long-range air operations. The Soviet Northern Fleet and the Northern Air Force were the pre-eminent Soviet commands; NATO responded by investing heavily in building proficiency at all levels of anti-submarine warfare while fighting in a contested battlespace on and above the waves. Crucial to this was defending the ultimate deterrent – the nuclear ballistic missile submarine (SSBN) forces assigned to the Alliance.

NATO's theatre-wide anti-submarine warfare response was spearheaded by US technology. The development of hydrophone arrays on the seabed alongside towed systems, the pre-positioning of maritime patrol aircraft with increasingly efficient sonar buoys and layers of submarine forces with sophisticated passive sonars generally enabled NATO to detect Soviet submarines.[5] Shadowing Russian submarines became a highly specialist task, and due to its position and the historic

---

[5] Thomas G Mahnken, *Technology and the American Way of War since 1945* (New York, NY: Columbia University Press, 2010), pp. 67–72.

maritime authority that the Royal Navy had built by success against the German Kriegsmarine U-boat fleet, the UK followed on behind the US as NATO leaders in tactical anti-submarine warfare.[6]

The strategic importance of the Atlantic to the Alliance resulted in many navies concentrating their primary capabilities in blue-water underwater warfare. However, it was the UK and US that spearheaded the expertise to challenge, and deter, Soviet forces by generating extraordinary prowess in this domain. Associated talents in areas such as under-ice submarine operations and prolonged tracking in the deep waters of the Greenland-Iceland-UK (GIUK) Gap by ship kept the operational advantage with NATO. Few of the almost daily interactions between these naval forces were made public, and in many cases even political leaders did not know or understand the detail. Commanders were given operational freedoms to use judgement – garnered through years of experience – to shadow and collect intelligence without constant recourse to the UK for approval. As a result, the tension in the North Atlantic never achieved the prominence in British political debate. UK submarines sailed into the Russian bastion on a regular basis, gaining critical acoustic intelligence on new submarines, surface ships and sonar signatures that proved fundamental to maintaining a competitive edge against Soviet units in the North Atlantic.[7]

Command structures in NATO from the late 1940s evolved frequently to reflect these circumstances. Two major NATO commanders quickly emerged, Supreme Allied Commanders for the Atlantic and for Europe (SACLANT and SACEUR), who exercised equal authority. SACEUR had a fortified and bounded land theatre to control, SACLANT had no such clarity; the vast Atlantic offered numerous opportunities for the Soviet adversary to operate with freedom. The major subordinate commander charged with commanding the East Atlantic, through which all Soviet forces would have to transit, was the British commander-in-chief who had responsibility for British command and NATO control, under SACLANT. The majority of the senior commanders in the Atlantic were therefore US or British: a factor largely due to the scale of available forces and their SSBN fleets.

As the Cold War unfolded, political and military discussion became increasingly focused on military operations in continental Europe, as the critical battle was perceived as the action to halt a Russian advance across Northwest Europe. The advent of a new AirLand Battle (ALB) strategy further displaced the North Atlantic from its previous central position as

---

[6] John Roberts, *Safeguarding the Nation: The Story of the Modern Royal Navy* (Barnsley: Seaforth Publishing, 2009).

[7] James Jinks and Peter Hennessy, *The Silent Deep: The Royal Navy Submarine Service since 1945* (London: Allen Lane, 2015).

one of the two primary conceptual efforts for NATO.[8] ALB recognised the scale, mass and speed with which Soviet forces could overwhelm central Europe. It heralded the arrival of emerging precision-guided munitions and associated strategies required to overcome this challenge. One implication was a reduction in the Alliance dependence on sea-transported resupply to stem any initial attack, since the accuracy of new weapons and a new plan (attacking second-echelon forces) would mean a significantly shortened conflict. ALB is now acknowledged by the US to represent its Second Offset Strategy.[9] Soviet doctrine responded with an increased prominence of the potential use of tactical nuclear weapons on the battlefield, further complicating the political aspects of any future land campaign. This further reduced the significance of the sea route, as military planners envisaged that the rapid escalation of any conflict with nuclear weapons would reach a conclusion well before resupply could reach Europe. The shift in the conceptualisation of warfare from sea to land marginalised the North Atlantic as a theatre of conflict and of navies more generally. Member states reduced investment in naval forces and increased spend on aircraft, precision weapons and intelligence systems. Even the central role of maritime forces in the 1982 Falklands War did little to alter this dynamic: NATO had started to think of war against the Warsaw Pact as a predominantly land- and air-based activity and the British Nott Review, highlighted by Malcolm Chalmers in Chapter II, codified this approach. In the late 1980s, the military balance in the Atlantic still favoured NATO; the Soviet bloc lacked the sheer number of platforms to truly dominate the ocean. The result of a combination of these factors meant that the land environment, in tactical and strategic terms, had begun to dominate military theory: a distinct contrast to NATO defence thinking in the 1970s.

Adaptation, 1991–2011

After 1991, Russian–Western relations improved. Western defence budgets were trimmed as the so-called 'peace dividend' – mischievously labelled by some as 'social disarmament' – was paid. Military focus shifted away from Europe and reinforcement towards elective, expeditionary operations. This, in turn, drove investment decisions that once again prioritised land and air forces over naval ones. Threats to sea lines of

---

[8] Robert R Leonhard, *The Art of Maneuver: Maneuver-Warfare Theory and Air Land Battle* (London: Presidio Press, 1995).
[9] Jesse Ellman et al., *Defense Acquisition Trends, 2015: Acquisition in the Era of Budgetary Constraints* (Washington, DC: Center for Strategic and International Studies, 2016), p. 7.

communication in these expeditionary environments were perceived as insignificant and extremely low risk. Small Royal Navy contributions to international interventions did little to reverse this: the liberation of Kuwait in 1991, operations in the Former Republic of Yugoslavia between 1994 and 1997, Kosovo in 1999 and Sierra Leone in 2000 all had maritime elements to them – some crucial – but gained little public or political attention.[10] This UK view was broadly representative of NATO's activity. Post 2001, both parties became increasingly engaged in counter-terrorist and counterinsurgency campaigns outside traditional NATO areas, specifically in Afghanistan and then Iraq. The UK national focus on these wars of choice resulted in continued decline in interest regarding maritime matters, leading some to declaim an era of 'sea-blindness'.[11]

The welcome need to embrace the newly liberated Eastern European states within the NATO community led to the creation of Allied Command Transformation, established at the expense of SACLANT. The Alliance focus on the Atlantic waned further and subordinate headquarters were aligned under SACEUR. Outside nuclear deterrence, maritime inputs were perceived as tactical in nature and spread across several smaller headquarters; they busied themselves with maritime security operations of varying levels of complexity. Forays into counter-piracy operations in the Indian Ocean from 2009 (Operation *Ocean Shield*) did little to counter the orthodoxy of modern military operations: wars were fought on land and were primarily fought about the people.[12] Only the US retained naval forces capable of conducting complex, integrated and joint warfighting operations. However, even the US Navy experienced resource constraints and, against a backdrop of healthy relations with Russia, prioritised spending in areas away from anti-submarine warfare. Across the board there was significant atrophy in experience and skill.

During the 1990s and the early part of the twenty-first century, future UK force designs and structures were established through analysis of a series of complex civilian-military scenarios in which Britain might find itself involved as a sole responder or as part of a coalition. The bulk centred on land operations providing outcomes such as peace enforcement, complex yet short-duration intervention or long-term peacekeeping. They were graded by 'scale of effort' and length of likely engagement and rarely

---

[10] Bradford A Lee and Karl F Walling (eds), *Strategic Logic and Political Rationality: Essays in Honor of Michael I. Handel* (London: Routledge, 2003), p. 279.
[11] Maria Fusaro, *Political Economies of Empire in the Early Modern Mediterranean: The Decline of Venice and the Rise of England, 1450–1700* (Cambridge: Cambridge University Press, 2015), p. 189.
[12] Rupert Smith, *The Utility of Force: The Art of War in the Modern World* (London: Penguin, 2006).

championed the necessity of credible deterrence in high-end warfare through NATO (it was assumed that the UK would never contribute more than 15 per cent to an Alliance force) in challenging regions such as the Atlantic. State-on-state conflict involving the UK or a NATO member was not envisaged as a likely future scenario that should drive the design of modern armed forces. If it did happen, it was essentially going to be an effort that relied on adequate warning and investment. This approach was mirrored across the Alliance and manifested itself in a preponderance of complex, expensive air warfare destroyers with limited additional utility (in the UK's case, the Type 45), large multirole platforms with amphibious capability and ocean patrol vessels. Specialist anti-submarine warfare platforms (ships and submarines) and modern maritime patrol aircraft struggled to gain recognition through these narrow, near-term force development scenarios and processes. The general-purpose vessel became the currency of choice in many states.

Russia also experienced resource challenges on an even greater scale; much of the Russian surface fleet was scrapped or mothballed. Nevertheless, resources continued to be found to keep a smaller submarine force modern and capable – essential to ensuring sovereign freedoms and to securing control of the Arctic and High North. Financial constraints forced Russian naval commanders to focus on people and technology, sharing experiences and building knowledge across their now-elite submariners while continuing to invest in missiles and submarine systems. In essence, the Russian navy developed a highly competent – arguably world-beating – seedcorn of submarine operators and boats, ready to be enlarged into a potent military capability when money became available.

The Royal Navy experienced a steady decline in hull numbers coinciding with a strategic shift to the Indian Ocean and the Gulf. The move to expeditionary capabilities, and a shift from deep-water (North Atlantic) skills to those needed for land attack, was codified by a paper called 'The Maritime Contribution to Joint Operations'.[13] The new-look Royal Navy was to be structured around a carrier and an amphibious task group designed to project power into areas of strategic interest. The hunter-killer submarine force, originally designed to search for and attack its Russian counterparts in the North Atlantic and Arctic, was rerolled for land attack with the inclusion of Tomahawk missiles in its arsenal; the force rarely forayed into the northern waters of the Atlantic. The *Invincible*-class aircraft carriers, traditionally anti-submarine warfare

---

[13] Geoffrey Till (ed.), *The Development of British Naval Thinking: Essays in Memory of Bryan Ranft* (London: Routledge, 2006), p. 189.

carriers, became platforms solely to provide close air support missions for land operations, eventually being decommissioned in 2010; the British anti-submarine warfare Striking Force was all but disbanded. The perceived end of the Russian threat consigned the North Atlantic to a minor role in the eyes of the British military establishment: the ability to send submarines under the Arctic ice was lost (albeit it is now in the early stages of reconstruction), maritime patrol aircraft were scrapped and while ships might have continued to be designed for an Atlantic climate, they rarely operated there and the Royal Navy largely disengaged from NATO's standing frigate forces. British military thinking had swung from a force designed to counter the Russian threat to a more generic structure. That shift included a diminution of the associated doctrine, skills and tactics for anti-submarine warfare other than those required for defence of the nuclear deterrent. The final nail in the coffin of British dominance in anti-submarine warfare was with the emergence of important maritime security operations serviced by single unit deployments across the globe, which exacerbated the withering of underwater warfare competencies.

Contemporary Comparisons

It is clear that Russian philosophy had changed, with campaigns in Chechnya (2001–04) and Georgia (in 2008) indicating a distinct, longer-term shift in outlook and appetite for its own rather aggressive interventions. During this period, the British military high command was focused on activities in Iraq, Afghanistan and Libya: it reassigned much of the Russian section of intelligence analysis in Defence Intelligence, preferring thematic analysis of future trends and threats in the Middle East. Procurement programmes continued their post-Cold War focus on small numbers of boutique and costly platforms designed for expeditionary warfare against less technologically capable adversaries. Russia's strategic shift in building new submarines and ships, embracing advanced technologies, rationalising command and control structures, and accelerating military activity in the Arctic and High North may well have been underestimated in the West.

Russia has always viewed NATO as a threat to its national security, and especially so following the expansion of the Alliance. Therefore, it has not been surprising to see renewed Russian activity in the Atlantic. No longer simply seeking to demonstrate an ability to interdict and disrupt transatlantic trade, Russian naval forays have become highly politicised attempts to embarrass UK, US and NATO forces. When two Russian *Akula*-class (Russian designation Project 971 Shchuka-B) submarines undertook patrols off the US eastern seaboard in 2009 and again in 2012, it even made it into the popular press.[14] This was part of increasingly frequent missions by the Russian submarine fleet into the North Atlantic and the number of patrols continues to rise, often coupled with air patrols and sorties by long-range bombers. This is likely to continue. Moscow is challenging NATO to react, hoping to sow the seeds of disunity and to demonstrate that it is an Alliance lacking cohesion, coherence and intent. To Moscow, the North Atlantic is a key region to demonstrate this dynamic.

The scale of Russian forces in the land, sea and air domains is not what it was during the Cold War and commentators often mistake rhetoric from Moscow as genuine military capability. Russia lacks the scale of forces to seriously challenge the West, especially at sea.[15] Mitigating this factor, former Soviet maritime doctrine (emphasising extended effort over devastating speed) has been replaced by concepts that favour dynamic and comprehensive warfare – exploiting vulnerabilities and seeking opportunistic victories. NATO's response must equally morph into a more coherent plan to counter Russian strengths if it is to have value. Moscow's contemporary strategy is more a mix of threshold warfare and blitzkrieg: a coherent effort to distract, disrupt and confuse individual states while working on a wider strategy to undermine NATO itself. At sea, this might mean the use of the Atlantic to mask and complicate decision-making in coincident activity within the Baltic Sea or in the approaches to the Mediterranean. For such a military strategy to succeed, it is essential that the planned US resupply and reinforcement of NATO be held at credible risk by Russian naval and air forces. Access and persistent presence in the North Atlantic is key to such a plan.

In the Atlantic, Russia has fewer operational submarines but NATO also has fewer ships and, critically, far fewer anti-submarine warfare assets (aircraft, ships and submarines) to respond. As evidenced by priorities placed on anti-submarine warfare from 2011, naval leaders in the US, France and the UK recognised the threat from around 2010 onwards; however, there was little evidence of any realignment of long-term resources or investment. The 2016 Warsaw Summit declaration recognised the increasing importance of the maritime domain,[16] but the North Atlantic as a geographic region received only one mention in the text, despite an agreement that seemed to invest much in the need to deter Russia.[17]

---

[14] Mark Mazzetti and Thom Shanker, 'Russian Subs Patrolling Off East Coast of U.S.', *New York Times*, 4 August 2009.
[15] International Institute of Strategic Studies (IISS), *The Military Balance 2017* (London: Routledge, 2017).
[16] NATO, 'Landmark NATO Summit in Warsaw Draws to a Close', 9 July 2016.
[17] NATO, 'Warsaw Summit Communiqué', (2016) 100, press release, 9 July 2016, <http://www.nato.int/cps/en/natohq/official_texts_133169.htm>, accessed 27 January 2017.

Thus, UK military commanders must expect Russia to continue to exert influence through the North Atlantic by distraction, confusion and strategic *maskirovka*,[18] and for this to increase in frequency, duration and breadth of military activities in the future. A strategic link between Russian military activities in the Baltic, the Atlantic and the Arctic is likely, and connectivity will be well established in Moscow where an asymmetric approach will continue to form a crucial part of operational design. Russian maritime and air forces consistently push NATO commanders to the limit of their authorities: the postures of Russian submarines, warships, surveillance vessels and aircraft during the recent deployment of the sole Russian aircraft carrier, *Admiral Kuznetsov*, to Syria are worthy examples here.

The balance of power in the Atlantic is different to that expected at the end of the Cold War yet, while NATO's plans for the Baltic are a very welcome step forward, the Alliance still needs to demonstrate credible deterrence more widely. An Alliance plan for dominating the maritime Atlantic has not been refreshed. Herein lies the rub for the UK and NATO. The North Atlantic, while present in title for the Alliance, forms no part of conceptual, coherent planning. This shortfall must be addressed if coherent responses and worthwhile contingency plans are to be generated. Something has to change.

How to Face the Russian Challenge

The outcome of the Warsaw Summit in July 2016 was confident and progressive, sensibly building on the Wales outcomes and reflecting the evolutionary nature of the Alliance. The challenge is to translate clear political intent into tangible changes within the Atlantic. This does not demand disproportionate strategic prominence for the theatre; what it does mean is recognition of the vital battlespace that is the Atlantic and considered debate as to how the challenges posed by an assertive Russian navy are to be addressed.

Central to this discussion will be command and control, the means by which core competencies are restored, the opportunities presented by technology and the interplay between national and Alliance planning. In all of these the UK has a respected, authoritative voice which should help shape the debate and inject momentum to all those areas, especially as the most recent Strategic Defence and Security Review reiterated the centrality of NATO (reinforced by the British decision to leave the European Union) to UK security. The UK should therefore seek to restore its position of leadership in Atlantic operations.

---

[18] In Russian, *maskirovka* means 'masking', but in Western military doctrine this has a meaning associated with deception.

There are a variety of different ways available to the UK to respond to the threat posed in the North Atlantic, both from inside and outside NATO. A starting point could be to help to lead the discussion of how central the Atlantic is to current Alliance thinking. The community is bedevilled by assumptions, further complicated by misaligned thinking on just what should occur in the Atlantic in a time of crisis. It is often seen as ducking a hard question to blithely suggest a policy review or generating a concept for a specific area: a bureaucratic tool to deter essential action. Yet, in this case, a renewed NATO strategy on the Atlantic would be extremely helpful, particularly given the levels of Russian activity over the past two years that have so far elicited an incoherent NATO response even in terms of posture. The debate is needed and the UK should be bold enough to start it.

The demise of SACLANT with its primary focus on the Atlantic region, while entirely understandable, is still keenly felt. Today, maritime authority within the Alliance rests with Allied Maritime Command (MARCOM) in Northwood, the original home of CINCEASTLANT (Commander-in-Chief, Eastern Alliance). MARCOM is now the Alliance's sole maritime advocate, the trusted adviser to SACEUR and the champion of all issues pertaining to the global maritime domain, including assured delivery and execution of component parts of the Readiness Plans that would be led by one of the two Joint Force commanders – a very broad canvas to be championed by a small staff. Most notable among the other Alliance operational maritime commands is Striking and Support Forces NATO (SFN), which has key responsibilities for many issues, including the integration of US strike and amphibious assets into NATO operations. It also fulfils a variety of on-call command functions and has responsibility for several Atlantic contingency plans. And its role further includes potentially supporting Norway. SFN reports directly to SACEUR and, while relations with MARCOM are generally good, there are no formal procedures to champion Atlantic importance or regional connectivity.

Each NATO member state that borders the Atlantic has, quite rightly, national contingency plans. The evolution of maritime doctrine, concepts, future capabilities and warfighting development rests with individual nations, shepherded by Allied Command Transformation. However, much of this is stale. Many more aspects rest upon the Atlantic canvas – covert anti-submarine warfare plans for the protection of nuclear deterrent assets for instance. In the past, SACLANT held these together and drove coherence within the region and with wider NATO planning, especially concerning reinforcement and resupply. Today, it is problematic to identify where ownerships rests. It should be discharged through an expanded MARCOM that also needs additional authority and capacity to rebuild Alliance blue-water capability and to generate credible contingency plans. It should have the freedom to work with national authorities and, alongside other Joint Force commanders, tailor plans to ensure operational coherence with contributing nations. The Atlantic is critical to the integrity of large elements of Alliance outputs in Europe and thus should become a defined and primary output of that command.

As previously stated, in purely numerical terms, the collective maritime capabilities of NATO overmatch Russia by a considerable margin; however, not all those forces will be assigned to the Atlantic. Identifying a small cadre of countries to champion a renaissance in combined underwater warfare (that encompasses land, sea, air, space and cyber, which exploits technology and all arms of intelligence) would go some way to restoring much needed resilience. Clearly, should a crisis befall NATO, forces would be made available from contributing states. However, complex open-ocean warfare requires tenacity and the skills that can only be generated by regular exercises. For without this, naval commanders will struggle and Alliance-wide anti-submarine warfare skills are already perilously low. Partnership development, providing high-quality and repeated training while developing joint responses, should be championed. A united effort to rebuild this is required and exploiting MARCOM's knowledge and experience would be distinctly advantageous.

At the tactical level, NATO currently has five seagoing maritime commanders who are on call to lead the maritime element of contingent operations in a one-in-five-year cycle. These High Readiness Force (Maritime) (HRF(M)) Headquarters comprise deployable national commanders (and staffs), and are provided by the UK, US (through SFN), France, Italy and Spain and are the naval equivalent of Divisional Staffs for land forces, able to command anywhere in the NATO region and under any Joint Force commander. The trend in NATO, likely to be reiterated in the ongoing NATO command structure review, is for greater regional focus within the permanent commands so that understanding during times of crisis is intuitive. Likewise, it allows the nominated commander to foster relationships during periods of stability while tailoring NATO processes to accommodate local strengths. This trend could be adopted for the HRF(M) commands so that one is permanently configured to the Atlantic and the High North. This would generate command competency and excellence in anti-submarine warfare, while husbanding the integration of assets assigned to the theatre. This is another role that the UK could fulfil for NATO. An important indicator to the wider community would be in dispensing with the operationally neutral nomenclatures of NATO's standing naval forces and restoring the role and name of Standing Naval Force Atlantic as a permanent force.

Acknowledging the centrality of the Atlantic both geographically and philosophically for the Alliance would re-establish and ground NATO in some practical realities that have, until now, been missing from the command. Only by evolving the command structure can the Alliance regain ground, conceptual superiority and the military initiative. The UK has a key role to play in such a discussion. It could promote a new plan, re-emerge as a key state for the North Atlantic theatre and spearhead a resurgence of anti-submarine warfare skills across the military spectrum. This would harness space, cyber, electronic warfare, industry and land forces rather than being confined simply to navies and air forces. Shared investment, centralised command, simplified doctrine, concentrated forces, economy-of-force effort through coordination and orchestration by an expert staff across the Alliance would be a legitimate signal that NATO (through MARCOM) and the UK will retake the initiative as a superior force in the North Atlantic.

Technology eventually gave the Allies the operational edge in the Second World War and technology helped the Alliance to keep ahead during the Cold War. It will do so in future. Be it in synthetic training or in acoustic arrays, autonomous underwater vehicles and satellite capabilities, more can be done at the national and Alliance-wide level. Technology will not in itself compensate for depleted assets; however, if used wisely it will help to magnify the outputs of those available today. This is mirrored by investment in the intellectual capital of navies. It used to be said, somewhat disparagingly, that ASW – the acronym for anti-submarine warfare – stood for 'Awfully Slow Warfare' due to the attritional nature of the battle. However, successful execution actually requires deep analytical skills, profound understanding of the battlespace and shrewd judgement. Does all of this have to rest solely in a single ship or aircraft? More of the analytical work can be delivered from shore facilities through remote working processes that would exploit the scarce anti-submarine warfare talent pool. Imagination is required.

Resources will remain finite with demand invariably outstripping financial resources – choice remains complex and often contentious. By investing in smaller systems across wider numbers of platforms, tasks can be shared more broadly. For instance, the UK has elected to fit the Tomahawk Land Attack Missile solely in the nuclear submarine fleet: nationally, it is regarded as a coercive weapon of strategic importance, requiring these enormously capable assets often to be held for long periods in a missile engagement box, at high readiness, in order to be ready to fire. This is hugely inefficient and could be equally well done by surface ships. For the UK, the Type 45 destroyers were built with sufficient space in the hull to carry the missiles and yet have never been given the capability. Addressing this would allow expensive submarines (of which the UK has too few) to return to their primary role as the *primus inter pares* in underwater warfare, rebuilding specialist skills and once again regaining mastery in the Atlantic and High North.

Realistically, the number of states that could make meaningful contributions to operations in the North Atlantic is quite small. Just as historically the UK and US held the lion's share of responsibilities for the North Atlantic, and mentored other navies in the development of anti-submarine warfare skills, a new larger group could help other states deliver more capability, more quickly. Such a modern group might include the UK, the US, France, Canada and Norway, and generate momentum across NATO members in their desire to rebuild blue-water capability. Such focused activity could provide benefits not just to bordering states such as Portugal and the Netherlands, but also to those which might be considering anti-submarine warfare skills – for example, Turkey, Greece and Italy.

If those states formed the nucleus of such a group, they would muster a formidable array of platforms and skills that might be a plausible first responder to counter Russian activity. They would be the most natural core providers of any Standing Naval Force Atlantic, enabling the development of key competencies for other Alliance members unable to contribute on a more permanent basis.

The idea of harnessing the benefits of NATO in a revised approach to the North Atlantic already has some traction.[19] Yet, from a military perspective, bridging the gap between aspiration and reality will be challenging. New strategies are useful, but implementing them in the way intended to meet the strategic need requires ships, aircraft, submarines and people. In matching the resources to the task, some different approaches have been identified.[20] The key dynamic identified by other research is the exceptionally high costs and skill levels necessary to contribute effectively in underwater warfare, along with the lack of political recognition for making such investments. Meeting the challenge within NATO requires a deliberate plan that provides commanders with more assets in the North Atlantic.

Some have suggested that an innovative technology will arrive that solves the issue of underwater transparency. Although a belief in technological silver bullets is attractive, research indicates that the reality of swarmed underwater drones locating and tracking Russian *Yasen-* (*Severodvinsk-*) class submarines is more than 30 years away.[21] Radical changes to the possibility of submarine detection that technology may put within NATO's grasp will have a profound impact on warfare in the Atlantic and beyond. Much as discussion on cyber has altered the

---

[19] James Foggo III and Alarik Fritz, 'Fourth Battle of the Atlantic', *Proceedings Magazine* (Vol. 142/6/1,360, June 2016).

[20] Peter Roberts, 'European Sea Power', in *The Handbook of European Armed Forces* (London: Oxford University Press, 2016).

[21] Bryan Clark, 'Game Changers – Undersea Warfare', statement before the House Armed Services Sea-Power and Projection Forces Subcommittee, 27 October 2015.

fundamental dynamics of war for many, so too could the arrival of an operational hyper-spectral sensor array[22] – with consequences for strategic decision-making underpinned by classical concepts such as assured strike and response. So while technological developments are possible, singular reliance on them as an effective counter to Russia in the North Atlantic would be a courageous move and at the same time highly unpredictable in terms of second- and third-order effects.

Alternatively, NATO states might make more national military assets available for missions in the North Atlantic. Few states, however, possess vessels suitable for such a mission. The requirement for the persistent presence of quiet, anti-submarine warfare-centric platforms limits the utility of many potential contributions from European states. Light frigates, amphibious ships, aircraft carriers, conventionally powered submarines, electro-optic-centric drones and platforms reliant on constant connectivity are of little benefit to commanders in the North Atlantic. Long-endurance, self-sufficient, independent and passive anti-submarine warfare-capable platforms are the necessary tools for the environment and these are in short supply. Where they are possessed by states, their capabilities make them highly desirable assets for longer deployments in other theatres. There has also been a trend since 2001 for these larger platforms to have less specialist anti-submarine warfare equipment, further reducing the potential pool of resources available for operations in the North Atlantic. The probability of this approach being successful is therefore small.

During the Cold War, some NATO member states were persuaded to invest in specialist areas of naval warfare. Greater specialisation at national levels allowed state investment to be maximised and generated true expertise in warfare areas available to all NATO members. This pattern of specialisation fell away after 2001, and states invested in broader military capabilities that could contribute to national influence across the spectrum of conflict. More states built coastal protection vessels instead of frigates and the global fleet of maritime patrol aircraft fell markedly. Given the national industrial strategies being advanced in many NATO states, greater national specialisation in warfare areas appears unlikely to bear fruit, even if it were to become an accepted strategy for the Alliance. Few states would sacrifice national independent military capabilities and become dependent on the political and public will of a partner.

Some of those interviewed believe that it is the cost implications of investing in North Atlantic capabilities that prevents many states in the Alliance from developing such skills or contributing more broadly. NATO

---

[22] David Stein et al., 'Hyperspectral Imaging for Intelligence, Surveillance, and Reconnaissance', Space and Naval Warfare Systems Center Pacific, 2001, <http://www.dtic.mil/dtic/tr/fulltext/u2/a434124.pdf>, accessed 30 January 2017.

has previously tackled this by developing a NATO-owned fleet of specialist, high-cost platforms that were manned by personnel from across contributing states. The NATO E3 airborne early warning (and command and control) aircraft fleet continues to demonstrate the viability of this approach some 25 years after its formation. It has seen deployment in almost every conflict and campaign that the Alliance has conducted since 1996. A combined, centrally owned force of expensive, deeply specialised assets operated by personnel from across the Alliance could mitigate some of the capability gaps in the Alliance, for example in the development of a NATO maritime patrol aircraft fleet.

Yet even these options for increasing the forces available in the North Atlantic miss the more seminal issue for NATO. The current separation of NATO areas breaks most of the principles of war as acknowledged by British, American, French, Prussian, Chinese and even Russian schools of military theory.[23] Boundaries and seams between NATO's geographic areas drive inefficiencies and have incoherent Graduated Response Plans assigned to them. Such historic domains no longer suit or represent vital national interests for member states and need a radical overhaul along with the command structures and relationships that accompany them. The Atlantic needs to be recognised as a theatre in its own right again, possibly at the same level of importance as the European theatre. The mission of such a headquarters, logically an emboldened and reinforced Joint Commander building in Northwood, would not be securing the North Atlantic for reinforcement. However, it would counter Russian activity across the whole of the Alliance's northern domain, and link to bodies and strategies in the Arctic and High North. Rather than preparing for a long-term blockade of Russia (a rehash of Kennan's containment strategy of 1947), it might perhaps be planning for a worst-case decisive battle with Russian forces or for deterring nuclear escalation by Moscow where, in the event of increasing tension, it sees no alternative strategies.

Conclusion

Britain and its military forces have historically maintained a significant interest in the North Atlantic in order to ensure freedom of trade, prosperity and connection to its empire, and continued to do so during the Cold War. The central role that the Royal Navy and Royal Air Force played there declined markedly from 1990 as the focus and doctrine of the UK armed forces shifted from countering the Russian threat to power projection beyond Europe's borders. Policy, doctrine and strategy since

---

[23] Jan Angstrom and J J Widen, *Contemporary Military Theory: The Dynamics of War* (Abingdon: Routledge, 2015).

2001 have continued to emphasise expeditionary operations as the driving force behind British military procurement and platforms, and have culminated in the UK concept of a Joint Expeditionary Force. This grouping of British military capabilities, underpinned by enabling networks and logistics, is constructed for fast, short-duration engagements on land. It is distinctly light on deep-water maritime capabilities, matched by a paucity of doctrine and concepts for fighting at sea. In short, the UK armed forces are simply no longer resourced or designed to counter a challenge from Russia, specifically in the North Atlantic. The shift in Royal Navy platforms away from underwater warfare towards a single expeditionary task group is less useful and relevant to the contemporary challenge of persistent, deep-water submarine hunting in the UK's backyard. Even if the political and military triggers to change that force design existed, it could not be achieved swiftly. Therefore, improving command and control, training, regional leadership and advocacy alongside emboldened partnerships is crucial. The UK cannot counter the Russian challenge alone. However, it can galvanise Alliance attitudes to the region and drive a more assertive strategy for this crucial theatre. It is a challenge worth confronting.

VI. THE UNITED STATES, THE NORTH ATLANTIC AND MARITIME HYBRID WARFARE

JAMES STAVRIDIS

An explosion below the water line suddenly rocks a giant container ship passing through the Dover Strait. The ship lists but remains afloat. Over the next few days, more explosions occur throughout the straits, seemingly at random. Crude waterborne mines are found free-floating in the strait, but the number of mines still adrift remains unknown. Forensics are inconclusive and the Kremlin denies any involvement. Shipping is rerouted, insurance premiums soar and millions of euros' worth of stock value disappear overnight. Politicians struggle to assure their publics right before critical elections.

Out of the darkness come fast attack boats headed towards an oil rig in the North Sea. They launch rocket and small arms attacks. The attacks are driven off, but not before the attackers manage to place an improvised explosive device (IED) near a pylon. After the explosion, the oil rig is out of commission for months and an obscure environmental terrorist organisation claims responsibility. The oil market is upended and gas prices soar.

In the middle of a busy trading day, the internet is suddenly paralysed by a break in the undersea cables carrying data to and from Europe and the United States. Financial markets are unable to function for hours until the internet can fully reroute traffic and the stock market plummets. The exact cause of the break in the cables is never known, but many suspect Russian interference.

These three vignettes are not outlandish scenarios by any stretch of the imagination. The North Atlantic is a target-rich environment for such events: congested shipping routes and harbours, oil rigs, underwater telecommunications cables and a host of vital infrastructure in the North Atlantic are crucial to the economies of Europe. Some $4-trillion worth of trade transits the Atlantic Ocean; the United States and Europe collectively are each other's largest trading partners. The North Atlantic has sea lines of communication (SLOC) and strategic economic routes on which the United States and the rest of NATO depend. Since the First World War, the US Navy has focused on the protection of North Atlantic choke points, and the control of SLOC has been almost synonymous with the US notion of collective national defence.

The new security landscape requires that NATO forces be prepared to defend these SLOC from potential Russian aggression. As the preceding chapters have demonstrated, a key part of Russia's strategy is to deny NATO access to land and sea areas bordering its territory. Russia has displayed this strategy in all regions of Europe. The Northern Fleet and the 'bastion' defence concept stand out as presenting a *strategic* challenge to the link between North America and Europe. NATO must understand that because the North Atlantic plays an important part in Russian military strategic calculations – as evident in its growing naval and air force laydown – it is essential that the region become more central to NATO's own planning, deployments and preparations. This requires an updated maritime strategy and a solid command and control structure specifically designed to deal with threats to the North Atlantic. It also necessitates updated intelligence and contingency plans for the region. NATO must invest in the high-intensity blue-water capabilities that it needs to conduct realistic and extensive training and exercises. Additionally, NATO members and partners must look beyond the traditional SLOC scenarios because in today's globalised economy, the very infrastructure of the North Atlantic has become vulnerable to maritime hybrid warfare – the subject of this chapter.

Hybrid warfare, although not a Russian construct per se, is a concept that the Kremlin understands well and will exploit unless the United States and the other members of the Alliance are ready to counter it. Russia has been exploiting asymmetric means and technological changes since the end of the Cold War to keep NATO off balance. Russia is skilled in hybrid and non-traditional approaches and has used them to turn its weak hand into a strong one. The Kremlin surely seeks to do so in all ways possible. Russia's adoption of hybrid tactics on land, as in Georgia and Crimea, clearly shows that expansion of such tactics into the maritime domain is not just a realistic possibility, but likely.

NATO must prepare now to deal with Russian hybrid warfare at sea before it is too late. Fortunately, this does not present an insurmountable challenge if the Alliance turns its attention to addressing it. Even today, the US Navy and NATO Allies and partners can defeat hybrid warfare threats if they prepare themselves properly. US strengths are well known: the US Navy can meet the high-end challenges in the North Atlantic through its partnerships with other NATO navies, with the Allies demonstrating collective resolve and readiness and rapidly adopting new strategies. NATO can adapt its existing strengths in these areas to overcome hybrid challenges as well. To do so, NATO members and partners must understand what maritime hybrid warfare is, how it might manifest itself in the North Atlantic and how to counter it.[1]

What is Maritime Hybrid Warfare?

Much has been written about the emergence of hybrid warfare in a variety of global scenarios, notably in the Russian invasion of Ukraine and annexation of Crimea. To date, this type of warfare has largely been confined to land operations in terms of both actual practice and theoretical discussion, but the emergence of maritime hybrid warfare seems inevitable. Now is the time for the US Navy to begin thinking about these scenarios and how to respond to them, both for its own forces and on behalf of NATO Allies and partners.

As this author has noted previously, the broadly agreed-upon tenets of hybrid warfare ashore include eight key elements that NATO members must understand both individually and collectively. First is the creation of real strategic effect at the tactical level, sometimes referred to as the impact of the 'strategic corporal'; second, the deployment of 'soldiers' in unmarked uniforms – sometimes referred to as 'little green men' – making their status ambiguous under international law; third, the dissemination of false and highly inflammatory rumours; fourth, a strong presence on social networks, generating propaganda and lies; fifth, special operators acting across the entire spectrum of violence, from pinprick attacks to larger assaults against several high-value infrastructure, commercial or military targets; sixth, the use of insurgent techniques – including car bombs, torture and kidnappings – to intimidate the population; seventh, the incorporation of non-military forces such as police into military operations; and, finally, the use of sophisticated cyber campaigns.

This combination of activities has remained effective in a variety of scenarios in so-called grey zones of conflict. Russia's actions in Ukraine perfectly reflect the way such conflicts combine the different elements. For example, Russia very capably used 'little green men' during the invasions and the subsequent illegal annexation of Crimea, which allowed

---

[1] This article is adapted from a similar piece published by Admiral Stavridis. See James Stavridis, 'Maritime Hybrid Warfare is Coming', *Proceedings Magazine* (Vol. 142/12/1,366, December 2016).

the Kremlin to spend weeks denying the presence of any Russian 'troops' on Ukrainian soil.

The fundamental goal of hybrid warfare is to find the space short of obvious military action that nevertheless has direct and recognisable tactical, operational and strategic impact, and to compress hostile activities into a zone characterised by sufficient ambiguity to give an aggressor a better chance of accomplishing an objective without full-blown, overt offensive action.

Hybrid warfare has significant general advantages likely to be particularly attractive to the Kremlin. It can allow a state to conduct operations to intimidate, degrade and destroy an opponent's capabilities without certain attribution, which permits greater latitude of activity and reduces the extent of criticism and the likelihood of sanctions imposed by the international community. It can bestow the advantage of surprise, as a recipient may not suspect the punch that is about to land. Its techniques give the user effective control of the tempo and timeline of events, thanks to the inherent ambiguity of the actions being undertaken. It is much less expensive than building the massive and capital-intensive platforms needed to conduct conventional littoral warfare.

So far Russian hybrid warfare has apparently focused on land-based operations, but the West must prepare itself for Russia's possibly adapting its expertise in hybrid warfare to the North Atlantic. Russia might contemplate doing so for many reasons. Executed well, it could negate or reduce the effectiveness of high-end US maritime capabilities and weaken the Alliance by eroding the economic strength of its members, as well as public trust in NATO military capabilities and the credibility of elected governments. It could also raise Russia's prestige, especially in the eyes of Russian populations, both at home and abroad.

Russian hybrid maritime warfare would likely have two primary characteristics. First, it would probably draw on the full range of capabilities of the Russian armed forces. Second, to appear somewhat ambiguous to outside observers, it might feature the use of civilian vessels (tramp steamers, large fishing vessels, light coastal tankers, small fast craft, and even 'low slow' skiffs with outboard engines), third-country flagged vessels and even unmarked craft.

Operations might be conducted or commanded and controlled by Russian military personnel out of uniform – 'little blue sailors' – similar to the 'little green men' who are not recognisable as regular, uniformed personnel. To give the appearance of non-state action, these actors might be categorised as nationalists, rogue actors, terrorists or even 'vacationing' sailors acting on their own volition. Such 'little blue sailors' would not have any markings on their clothing, would not carry passports and, if captured, would deny being part of any organised military force.

Maritime hybrid warfare platforms could carry a variety of weapons, from light arms to heavier calibre but temporarily mounted machine guns, hand-held surface-to-surface missiles and light surface-to-air missiles. 'Sailors' might also have access to effective, privately available equipment such as drones, high-intensity laser dazzlers, sound emitters, tear gas dispensers, water cannons and other non-lethal weapons. Their command and control would be compact, civilianised and largely composed of off-the-shelf systems. However, the forces would have the ability to deploy overhead unmanned sensors (which would be light, smart, inexpensive and disposable). They might even have the ability to deploy sonobuoys and underwater and surface unmanned sensors and to emplace permanent sensor nodes on the seabed. All of this technology could be maintained ashore by Russian special forces.

At the most sophisticated level, Russian forces could conceivably employ 'Q-ships', specially designed to look like coastal steamers or other small-to-mid-size commercial vessels, but with concealed ports built into their sides for weapons and the ability to launch speedboats and even missiles from internal bays. These Q-ships could function as motherships for even smaller and less sophisticated vessels. Such vessels could also surreptitiously discharge mines closely resembling maritime IEDs and manufactured in a crude and untraceable fashion.

With all this in mind, and given Russian military capabilities, the Kremlin could employ a wide range of methods to achieve strategic impact with sufficient ambiguity to avoid being fully recognised as culpable. Surreptitious discharge of free-floating 'homemade' mines in strategic ports, harbours or canals could threaten and even shut down shipping traffic. Fast-boat or waterborne IED attacks on North Sea oil platforms by 'environmental terrorists' would have significant impact on the global oil market and produce a chilling effect on further investment, given the costs of remedying environmental effects. Deniable undersea attacks on ocean-bed telecommunications cables or shore terminals would threaten global information flows and financial networks. 'Pirate' attacks by political terrorists on NATO-affiliated shipping or naval facilities would severely hamper Alliance operations. Mafia-style actions (such as kidnapping and car bombings) against coastal fishermen or fishing industries would destabilise economies and generate political pressure on national governments. The vulnerability of offshore oil terminals to seaward attack is well known. Lloyd's of London notes that:

> From a structural perspective, offshore terminals are vulnerable for a variety of reasons, and specifically in certain ways. Shore-located terminals can be threatened from landside, air, and seaward. Assuming a non-conventional conflict environment (e.g., major

interstate war), given that most fixed-installation security systems and protocols tend to be focused toward landside approaches, the greatest vulnerability is generally from the seaward axis.[2]

This concern is not hypothetical: experiences in Iraq demonstrated the ability to gravely damage or even destroy huge hydrocarbon installations in the Gulf in 2004.[3] Such attacks could even take place against undersea information cables.

Each of the above-mentioned attacks would have some form of strategic effect on NATO countries, could be carried out with plausible deniability and could easily be supported by cyber attacks and propaganda efforts (such as sympathetic fake news). Depending on timing, they might also be used to support certain political candidates over others. All in all, from Russia's perspective, the benefits of maritime hybrid warfare would likely outweigh the costs. An age of maritime hybrid warfare is coming and the United States and its NATO Allies and partners need to prepare for it.

How Can NATO Prepare?

The United States and its NATO Allies and partners must start to consider their responses to hybrid warfare at sea. They may need to develop new tactics and technologies, work closely together and build their own hybrid capabilities to counter those deployed by Russia. The good news is that the core operational principles to which NATO already adheres can adapt to this growing threat. There are five principles to guide NATO responses to the challenge of hybrid warfare. First, to retain a strong and capable deterrent network of forces committed to regional security and regional stability, including against hybrid warfare threats; second, to maintain the Alliance and its unified and unshakeable commitment to Article V – even when faced with 'grey zone' incursions; third, to maintain a technological edge and positive balance of forces today and into the future, and adapt them to deliver better protection against hybrid threats; fourth, to demonstrate, through its strategic communications and actions, that NATO is a 'defensive' alliance committed to stability in Europe – and that it will not tolerate any threat, even hybrid ones, to that stability; fifth, to continue to operate, train and react to threats as a seamlessly connected partnership of navies that can, at will,

---

[2] Herbert-Burns et al. (eds), *Lloyd's MIU Handbook of Maritime Security* (Boca Raton, FL: Auerbach Publications, 2009), p. 154.

[3] See, for example, Commander, Joint Forces Maritime Component Commander/ Commander, US Naval Forces Central Command/Commander, US 5th Fleet Public Affairs, 'Two Sailors Killed in Arabian Gulf Oil Terminal Attacks', 24 April 2004; David E Sanger and Eric Schmitt, 'Russian Ships Near Data Cables are Too Close for U.S. Comfort', *New York Times*, 25 October 2015; Greg Miller, 'Undersea Internet Cables are Surprisingly Vulnerable', *Wired*, 29 October 2015.

interoperate wherever and whenever needed to meet any threats to its collective security, national sovereignty and the inviolability of the articles of the Alliance, and expand operations and training to include hybrid threats.

With these principles as guides, the way ahead for countering hybrid warfare attacks in the North Atlantic is clear. NATO should consider a series of action points. All are important, but collectively they amount to a 'comprehensive approach' to maritime hybrid warfare.

Build Intellectual Capital

The most important activities NATO can undertake immediately are to study, analyse and fully understand how the ideas of hybrid warfare as practised today will both translate into the maritime sphere and develop there in lethal ways. This means not only looking closely at the current practices of Russia and others, but also having its own strategists and tacticians focus on this possibility. NATO schools and tactical centres should incorporate hybrid warfare into their curricula, plans and analyses, and leverage national lessons learned.

Develop Tactical and Technological Counters

NATO should develop and support consortia and related initiatives to test and field new technologies that better protect critical infrastructure in the North Atlantic and provide counters to the weapons and techniques described above – from crude mines to armed speedboats, to cyber and psychological attacks. The Alliance should also identify those national capabilities and tactics already in use across NATO that can counter potential hybrid attacks.

Improve and Coordinate Information Sharing and Strategic Messaging

A significant aspect of defeating hybrid warfare includes intelligence and information sharing to deny adversaries the ambiguity they seek. NATO can accomplish this in part by linking national-level intelligence and interagency resources more closely and even leveraging private sector elements to help defend critical infrastructure, as well as observe and record suspicious activities. NATO members should also improve cooperation in and coordination of their strategic messaging and information to rapidly and effectively counter pro-Russia information operations.

Work with Partners and Stakeholder Organisations

While many NATO countries are potential targets of Russian maritime hybrid warfare, other states, such as Sweden and Finland, may also become the objects of such attacks. The Alliance should work with these countries to improve information sharing and joint efforts to defend critical infrastructure, as well as to counter Russian propaganda. NATO should also use forums such as the International Seapower Symposium or national-level counterparts to encourage dialogue, exchange best practices and share intelligence on this emerging concern.

Train and Exercise Against Maritime Hybrid Warfare

NATO must begin to practise countering these types of operations, perhaps as part of the training cycles for national navies, surface, subsurface and aviation. The ambiguity of the scenarios described above will require education and training in rules of engagement, operating conventional systems against unconventional forces at sea and learning to act more like a network at sea in littoral areas. NATO has come a long way in understanding counterterrorism and networks, and could incorporate much of that knowledge into its training and exercise programmes. Furthermore, NATO could expand its larger exercises, such as *Baltops* and *Trident Juncture*, to include hybrid warfare scenarios.

Leverage Coast Guards and Law Enforcement

NATO states have exceptionally capable coast guards and law enforcement organisations. Many have already given thought to countering hybrid attacks and are enthusiastically doing so at the national level. Furthermore, many organisations in the private industrial sector (such as shipping companies and oil corporations) have clear incentives to defend against hybrid attacks. NATO should expand partnerships with regional coast guards, law enforcement and the private sector to enable and improve maritime security at critical locations. Many coast guards and coastal forces already have the technologies and tactics to counter maritime hybrid warfare techniques, and incorporating their experience into this tactical arena would prove highly beneficial.

Increase Steady-state Maritime Presence in the North Atlantic

NATO can increase its steady-state maritime presence in the North Atlantic through rotational deployments of ships, submarines and maritime patrol aircraft. Such presence would signal what NATO members and partners consider important to their security and familiarise sailors with these waters. It would also form part of the wider deterrence strategy and offer credible defence.

When executing hybrid warfare on land, the 'little green men' can find sanctuary amongst the population and/or in the local terrain between operations. By contrast, in the maritime domain 'little blue men' might conceal their identity and intent through deception and ruse, but when an incident occurs, defenders have a fleeting window in which to respond or to strike at the attackers before they can return to sanctuary – whether amongst white shipping, inside territorial waters or beneath the ocean surface.

The US-NATO-EU coalition counter-piracy campaign in the Gulf of Aden and Indian Ocean offers an example of a successful counter-maritime hybrid warfare effort. While the pirates were not nearly as sophisticated as the Russians would be, they applied some of the same operational concepts in that they used deception and ruse to conceal their identity and intent. Once an attack commenced and the actors became clear, counterpiracy forces had a narrow window in which to respond before the attack succeeded. The length of that response time was inversely proportional to steady-state force densities: the more forces (ships) distributed throughout the battlespace, the shorter the time required to defeat pirate attacks. The same principle applies in the North Atlantic to defeat Russian maritime hybrid warfare.

As Supreme Allied Commander Europe, this author found that leading NATO is like conducting an orchestra in which no musician has the same sheet of music as the others and all the instruments come from different cultures and musical traditions. Despite everyone's best intentions to play a lovely symphony, the music that emerges at times takes on a rather strange quality. But, like an orchestra, when NATO gets it right it can be very powerful. NATO needs to get the maritime hybrid warfare threat right.

Conclusion: 'Stronger Together'

The deep blue water that Leif Erikson and later Christopher Columbus sailed centuries ago – thus connecting the New World and the mother countries – is the lifeline of cooperation between the continents. The transatlantic bridge is not one of steel, mortar or bricks, but one of ideas and ideals founded on democracy, individual liberty and the rule of law as outlined in the North Atlantic Treaty of 1949. NATO members must yet again prepare themselves for high-intensity warfare in the North Atlantic, as this Whitehall Paper demonstrates, but they must also prepare for lower-intensity maritime hybrid warfare of various complexities. The leaders of the US and other NATO navies must adapt their strategies, postures and tactics to the full spectrum of operations if they are to ensure continued dominance in the North Atlantic and elsewhere.

Unilateral action usually leads to disappointment: instead, alliances, partnerships and friendships are essential. NATO must emphasise international, interagency and public–private cooperation in creating security in the twenty-first century. This is important not only for conducting high-intensity Article V operations, but also for tackling myriad maritime hybrid warfare scenarios. As the threat of hybrid warfare increases, so too do the relevance and scope of the Alliance. There is nothing fundamentally new about incorporating unconventional, and unacknowledged, forces on the battlefield in surprising ways to undermine conventional forces and obscure attribution. What is changing is the likelihood that Russia might use such forces to obtain important tactical, strategic and political advantages. Inevitably, hybrid maritime warfare will prove a formidable challenge if the Alliance has not thought through its uses and how to counter it. This does not constitute 'mission impossible' if NATO applies its intellectual powers to the fullest. If NATO members are to remain 'stronger together', the Alliance must face these new challenges head on and with one shared vision to ensure the continued security and economic vitality of the North Atlantic.

CONCLUSIONS AND RECOMMENDATIONS

JOHN ANDREAS OLSEN

NATO must constantly renew itself if it is to remain the world's most powerful political and military alliance. Currently, NATO's most pressing concern is that Russia is challenging the collective defence and the political cohesion of the West. As part of this strategy, the Russian Northern Fleet and the bastion concept clearly stand out as *strategic* challenges to transatlantic defence because they threaten the link between North America and Europe. If NATO does not have effective control of the North Atlantic, or at least the ability to deny Russia naval access to this maritime domain, Russia could block or disrupt US reinforcement to Europe.

NATO countries are far superior to Russia should they muster their combined economic and military resources in a unified effort. NATO has repeatedly demonstrated unity and resolve at its summits in Wales (2014) and Warsaw (2016). Several members have increased their defence budgets to meet new demands and NATO has responded to increased Russian military activity through forward-deployed ground forces in the Baltic region and Poland in particular.

This Whitehall Paper suggests that it is time to broaden NATO's strategic aperture to include the North Atlantic. The main conclusion emerging from this study is that NATO must return to these waters. The North Atlantic must yet again be recognised as an operational space in its own right as well as a continuous and interdependent transatlantic theatre of operation. There is no substitute for NATO being the *prime structural driver* for improving credible deterrence and capable defence, and within the Alliance the United States, the United Kingdom and Norway have a special role to play in strengthening the defence of NATO's northern flank due to their geographic locations and capabilities. Although the defence of the North Atlantic and its maritime flanks must emphasise maritime forces, it must be a truly full-spectrum (joint) effort. The following paragraphs offer short recommendations for strengthening NATO's maritime posture in the North Atlantic; they represent the common ground for the authors who contributed to this volume.

Renew NATO's Maritime Strategy

The Alliance must revise and update its maritime strategy. The current version does not reflect the dramatic changes in the security environment that have occurred since it was published in 2011, ranging from Russia's military actions in Ukraine to the re-emerging contest for maritime supremacy in the North Atlantic. NATO's new strategy should be a framework for a deeper understanding of the North Atlantic and its relevance to transatlantic security. The new strategy must acknowledge the necessity of managing the full spectrum of tasks, from peacetime reassurance to maritime hybrid warfare – including addressing the vulnerability of undersea cables across the North Atlantic – but emphasise deterrence and collective defence. The new strategy should address evolving capabilities and newly formed initiatives rather than seek a revolutionary change to operations in the maritime domain. The strategy must fully take into account that the transatlantic link is one of NATO's key strategic enablers for peace and prosperity and that the defence of NATO's northern region and maritime domain can benefit greatly from integrating operations from three key positions: the UK; Norway; and Iceland. A new maritime strategy must explain how a credible defence of the North Atlantic concerns all parts of Europe, not only the maritime countries.

Reintroduce Extensive Maritime Exercises and Sustained Presence

More extensive training and focused Article V exercises, founded on formalised collective contingency planning, are imperative to communicate cohesion, strength and determination and thus to achieve effective defence and deterrence. NATO's military headquarters should be given a more prominent role in the exercise activities undertaken in the region. Members should link national exercises, including US exercises with major surface and sub-surface combatants, to NATO exercises. The High-visibility Exercise *Trident Juncture*, to be hosted by Norway in 2018, offers an excellent opportunity to test a comprehensive and combined effort and to take stock of contingency plans and capabilities. Maritime forces must train differently and must continually be evaluated by task force assessment teams to ensure quality, responsiveness and relevance. While extensive and realistic training and exercises are important, a sustained presence is crucial to demonstrate commitment and to ensure maritime situational awareness in the region.

Reform NATO's Command Structure

The current NATO command structure is not appropriate to countering present-day strategic challenges and requirements, as it was designed for out-of-area operations, not deterrence and collective defence. NATO's credibility depends on a command structure that can lead large joint and multinational forces to defend NATO's territorial integrity, in addition to tasks such as crisis management, cooperative security and maritime security. A reformed command structure should reintroduce geographic areas of responsibility and ensure strong links with relevant national operational headquarters. One solution consists of increasingly double-hatting selected national headquarters to carry out specific tasks or activities during peacetime, crisis and war. Rather than invest in new infrastructure, the first step should be to enhance the existing Maritime Command (MARCOM) in Northwood (North London) and link it with US maritime competences. Once charged with the defence of the Atlantic, this improved command could also serve as a hub for military consultation among those states that would be most affected by, or involved in, Allied military planning and operations in this maritime domain. An enhanced MARCOM, a 'North Atlantic hub' for multinational, full-spectrum efforts, would require additional authority and capacity to rebuild the Alliance's capability to generate credible plans for blue-water operations, and include joint operational plans for surrounding countries.

Invest in Maritime Capabilities and Situational Awareness

A strong maritime posture in the North Atlantic depends on high-end, tailored warfighting capabilities. Since Russia is deploying long-range precision weapons systems, NATO's ability to deploy forces to exposed areas might be seriously restrained. NATO needs interoperable naval forces capable of establishing and maintaining all-domain access at short notice. NATO should develop a coherent and combined approach to anti-surface and anti-submarine warfare operations for the North Atlantic that harnesses and exploits not only naval and air forces, but also expertise from land forces and domain knowledge in cyber, electronic warfare, aerospace, technology, industry and intelligence. Maintaining the technological edge is critical. More forces must also be prepared and available for the battle of the seas and littorals, which would require both reinvigorating the standing naval forces and providing for follow-on response forces. Norway, the United Kingdom and the United States are about to expand their intelligence, surveillance and reconnaissance capabilities for operations in the maritime domain and in the northern region. This presents an opportunity for other NATO members to join that effort. Intelligence and surveillance is the first line of defence. Improved situational awareness and information sharing are vital to maintain knowledge about Russia's intentions and capabilities.

Enhance Maritime Partnerships

Since Russia is on the verge of establishing long-range anti-access capabilities that could affect all parts of Europe, any security approach must include both the Baltic region and the Norwegian Sea. Cooperation with selected partners, such as Finland and Sweden in the Baltic Sea, strengthens the Alliance. These countries' presence, capabilities, intelligence and profound understanding of the northern region, and their relationships with Russia, play a central role in maintaining security in the North Atlantic. It is therefore critical that NATO continue to operate, train and exercise as seamlessly as possible with key partners through multinational and bilateral agreements. Interoperability is the key to the strength of partnership among like-minded nations. NATO members and partners must continue to ensure that they can operate as a single force when needed. To do so, NATO should build on US naval partnership concepts to unlock even greater potential for integration and cooperation among NATO forces.

Prepare for Maritime Hybrid Warfare

NATO must give priority to high-end collective defence and deterrence in the North Atlantic, but there are other low-end threats that must also be taken seriously. NATO needs to exercise various maritime hybrid warfare scenarios, including reactions to attacks on high-value targets at sea and under the sea. Such operations differ strongly from high-intensity blue-water operations, but could potentially have a paralysing effect. Taking preventive measures are just as important as preparing response options. To counter hybrid maritime threats and attacks in the North Atlantic, NATO should build intellectual capital, develop tactical and technological countermeasures, improve and coordinate information sharing and strategic messaging, work with partners and stakeholder organisations, train and exercise across the full spectrum of operations, leverage coast guards and law enforcement and increase steady-state maritime presence. If NATO members are to remain 'stronger together', the Alliance must take these threats seriously, identifying vulnerabilities and capabilities, to ensure the continued security and economic vitality of the North Atlantic.

A New Security Approach to the North Atlantic

NATO must relearn some of the maritime concepts that dominated the Cold War period, while acknowledging that the present security environment differs greatly from that of the earlier period. The listed measures offer a comprehensive and sustainable approach for maritime supremacy in the North Atlantic to protect national sovereignty and territory in accordance with international law. There is a growing consensus in NATO that it needs to move in this direction. MARCOM has taken several initiatives to strengthen the maritime posture in the North Atlantic and to improve deterrence and collective defence, but it needs more resources to meet the new security challenges.

The North Atlantic remains essential to the commerce and trade that sustain the economic prosperity of all countries sailing in these waters. It is vitally important that the Alliance remains resolute and that its members and partners continue to contribute their fair share in securing and protecting the transatlantic bridge. After all, the North Atlantic Ocean is NATO's lifeblood.